AF597312

PICTURE BOOK OF AUTHENTIC MID-VICTORIAN GAS LIGHTING FIXTURES

Mitchell, Vance & Co.

PICTURE BOOK OF AUTHENTIC MID-VICTORIAN GAS LIGHTING FIXTURES

A Reprint of the Historic Mitchell, Vance & Co. Catalog, ca. 1876, with Over 1000 Illustrations

With an Introduction by

DENYS PETER MYERS

DOVER PUBLICATIONS, INC., NEW YORK

The publisher gratefully acknowledges the cooperation of Mrs. Renee Berman in making the reproduction of this book possible.

Published in Canada by General Publishing Company, Ltd., 30 Lesmill Road, Don Mills, Toronto, Ontario.
Published in the United Kingdom of Constable and Company, Ltd., 10 Orange Street, London WC2H 7EG.

This Dover edition, first published in 1984, is a republication of the catalog originally issued by Mitchell, Vance & Co., ca. 1876. Those plates which were tinted in the original edition appear here in black and white. A full-color plate has been omitted. A new introduction has been written for this edition by Denys Peter Myers.

DOVER *Pictorial Archive* SERIES

Manufactured in the United States of America
Dover Publications, Inc., 31 East 2nd Street, Mineola, N.Y. 11501

Library of Congress Cataloging in Publication Data

Mitchell, Vance & Co.
Picture book of authentic mid-Victorian gas lighting fixtures.

(Dover pictorial archive series)
1. Gas light fixtures, Victorian—Pictorial works. 2. Art metal-work—History—19th century—Pictorial works. 3. Gas light fixtures—Catalogs. 4. Mitchell, Vance & Co.—Catalogs. I. Myers, Denys Peter, 1916– II. Title. III. Series.
NK8438.M57 1984 749'.63'097471 83-25490
ISBN 0-486-24640-X

Introduction to the Dover Edition

This catalog of gas fixtures and fittings issued by Mitchell, Vance & Company at an undetermined date between 1870 and 1877 (probably in 1876), while they were at 597 Broadway, illustrates almost the complete range of their production. Only their line of "crystal glass" fixtures, illustrated by thirty lithographs in a separate catalog, was omitted from this series of plates. The firm was then at the zenith of its success, although it retained a considerable share of the lighting-fixture trade for many years thereafter. During the 1870s, its closest rivals were Cornelius & Sons in Philadelphia and Archer & Pancoast in New York, over both of which Mitchell, Vance & Company triumphed in the awards won at the 1876 Centennial Exposition in Philadelphia.

After a few sporadic and tentative experiments, gas lighting got off to an early start in the United States with the establishment of a chartered gas company in Baltimore in 1817. Boston followed in 1822, New York in 1823, and New Orleans in 1835. By 1840 there were eleven gas companies chartered in America, and by the last day of 1850 fifty-one companies had received their charters. A tremendous surge of the "go-ahead" spirit of the age occurred during the next decade, and by June 15, 1863, there were 433 gas companies doing business in the United States as well as twenty-three in Canada. By the Centennial Year, 1876, it could be boasted that every American settled community larger than a village had its own gas company. Furthermore, many country estates in places too small to support a gas company had private gas plants. The gas companies were, however, by no means evenly distributed through the various sections of the country. Of the 433 companies listed in mid-1863, most were in the well-populated and already industrialized Northeast. For example, Arkansas had only one gas company, while New York had eighty-four. Almost all of the companies manufactured coal gas. A few made "rosin gas" from wood, and only one (in Fredonia, New York) used natural gas.

A promotional brochure probably issued in 1850 pleaded the cause of gas in the following words:

> The advantage of gas-lights is manifest. It is much cheaper, compared with the light it affords, than any other. It saves a deal of time and labor, which would otherwise be expended in filling and trimming lamps, cleaning candlesticks and snuffing candles—to say nothing of the constant trouble and anxiety given by these operations, and the spots of grease and oil which follow them. Gas-lights are the very perfection of cleanliness. They can be fixed in any situations—and by means of moveable pipes, may be raised, lowered and transferred, according to choice or necessity. The light is agreeable and if properly managed, which management requires no trouble, gives no smoke. In point of cost proportioned to its brilliance, it is nearly one-third that of lard, burned in our solar lamps, and at least one-sixth that of tallow candles. (Archer & Warner. *A Familiar Treatise on Candles, Lamps and Gas Lights. . . .* Philadelphia [1850?].)

If the tremendous increase in the number of gas companies established during the 1850s is any indication, those arguments were effective in spite of the possibility—mentioned in the brochure—of explosions caused by the neglect of leaks.

Until 1836, when the Philadelphia Gas Company started operations, the comparatively few gas fixtures required in America were imported. In that year, the firm of Cornelius & Son in that city commenced making gas fixtures, most of them highly ornamental, and by 1857 the firm, which had become Cornelius & Baker in 1851, was credited with having manufactured nearly half the gas fixtures by then produced in the United States. By 1848 the Archer & Warner firm was also contributing significantly to the pre-Civil War preeminence of Philadelphia in the manufacture of gas fixtures. At the same time Henry N. Hooper & Company of Boston was producing work that equaled Cornelius & Baker's best, and the New York firm of Starr, Fellows & Company, which became Fellows, Hoffman & Company in 1857, escalated the competition with the Philadelphia firms. It was during the same decade that John S. Mitchell and his partners entered the trade.

The firm of Mitchell, Vance & Company was founded in 1854 as Mitchell, Bailey & Company, a partnership composed of John S. Mitchell, John Bailey, Anson H. Colt and Samuel B. H. Vance, incorporated under Connecticut law but doing business at 85 John Street in New York City. By 1856 the new company already had a fashionable retail address at 526 Broadway and a factory on West 24th Street. In 1860 the firm was reorganized as Mitchell, Vance & Company. The partnership formed in 1860 was composed of John S. Mitchell, Samuel B. H. Vance, Aaron Benedict and Charles Benedict. In 1873 the copartnership, which was still incorporated in Connecticut, was dissolved and reincorporated under the laws of the State of New York.

John S. Mitchell died at his residence in Tarrytown, New York, on February 1, 1875, and was buried in Waterbury, Connecticut, which may have been his birthplace. Following Mitchell's death, Charles Benedict became president of the firm, with Samuel B. H. Vance serving as vice-president and Edgar M. Smith as secretary and treasurer. The officers were also trustees of the company, together with Edward A. Mitchell and Dennis C. Wilcox. Those were the five men in charge of Mitchell, Vance & Company when it achieved its triumph, surpassing all its competitors, at the Centennial Exposition at Philadelphia in 1876. Unfortunately, nothing is now known of them except their names.

The firm continued to flourish until after the turn of the century. Its illustrated full-page advertisement in *Trow's New York City Directory* for the year ending May 1, 1881, carried the following text:

> Mitchell, Vance & Co. Manufacturers of Gas Fixtures, Fine Clocks, and Bronzes. Highest Award and Medals at the Centennial Exhibition. Crystal, gilt, bronze, and decorated [i.e., polychromed] gas fixtures in the greatest variety at low prices. Special designs for churches, halls, hotels, dwellings, etc. 836 & 838 Broadway, New York.

The retail address of the firm was 620 Broadway in 1860 and it remained there until 1870. From 1870 until 1877 the store was at 597 Broadway (which establishes the date of this catalog within that span of time). In 1877 the retail outlet was moved to a new and larger building that had been especially designed for the firm on the west side of Broadway at 13th Street, running through to Mercer Street. There, at 836–838 Broadway, the store remained until 1915. Another and final retail listing appeared in 1915, when the store address was given as 294 Madison Avenue. The factory and foundry, which in its heyday occupied most of the block west of Tenth Avenue between West 24th and West 25th streets, was listed at various addresses within that block until 1933, when Mitchell, Vance & Company, by then called The Mitchell Vance Company, ceased business in the midst of the Great Depression.

The centennial year, 1876, was a banner year for the firm. Mitchell, Vance & Company was the only one in its field to be written up in *Frank Leslie's Illustrated Historical Register of the Centennial Exposition:*

> The exhibit of Messrs. Mitchell, Vance & Co., of 597 Broadway, New York, in the Main Building, included chandeliers, gas-fixtures, bronze ornaments and fine clocks. Our illustrations [a thirty-light crystal chandelier, seven-light slide library chandelier, crystal standard and ecclesiastical standard] represent specimens of their wares, and fairly display their attractive and artistic character. It is unquestionable that these exhibits were quite the handsomest in their line.

The official report of the judges to the United States Centennial Commission, in speaking of Group XIV Class 223 (Apparatus for Lighting: Gas-Fixtures, Lamps, Etc.) read in part as follows:

> Under all the requirements exacted in gas-fixtures the whole exhibit of Mitchell, Vance, & Co., of New York, most fully met the approval of the Judges. In the specialties of colored ecclesiastical ware and slide-lights, that of the Archer Pancoast Manufacturing Company was the best; in brass goods of the later designs and finish the productions of Baker, Arnold, & Co., of Philadelphia, were superior; while the crystal fixtures of Jas. Green & Nephew, of London, and the Mount Washington Glass-Works, of Massachusetts, stood unequaled.

The text of the report on awards was the longest of any in the gas fixture category:

> 102. Mitchell, Vance & Co., New York, N.Y., U.S.
> GAS FIXTURES AND ECCLESIASTICAL WARE.
> *Report.*—Commended for the following reasons:
>
> 1. This exhibit is of a large, complete, and varied character, of special excellence in design, workmanship, and finish, and is arranged with great taste and skill.
>
> 2. In gilt and polished brass gas fixtures the exhibit is of excellence in the wide variety of designs employed, its elegance and artistic character, and the high order of finish attained. In combinations of metal with porcelain or glass, rich effects have been here produced.
>
> 3. In steel-finished fixtures a novelty of beauty and durability is presented.
>
> 4. The "double-slide" extension light presents certain features of durability and regularity of motion that are of merit, while the arrangements for avoiding the heating and smoking of the rest of the chandelier are unique.
>
> 5. In bronze fixtures, both real and spelter, this exhibit is of excellence, as well in workmanship and finish as in chaste character and tastefulness of design.
>
> 6. In crystal gas fixtures the size of the integral parts, the integrity of the character of the goods as "crystal" (few wires or chains being used, the arms, etc., being solid crystal), the beauty and taste as well as novelty of the designs employed, and the excellence of the material used, give this part of the exhibit prominence and value.
>
> 7. In ecclesiastical ware, altar and sanctuary lights, candlesticks, coronas, chancel rails, etc., the several exhibits of the medieval and Gothic orders are of high merit.
>
> The bronze and brass railings for church use are of excellence and beauty, being architecturally correct in their respective schools. (United States Centennial Commission. *International Exhibition, 1876. Reports and Awards* Group XIV . . . . Philadelphia: J. B. Lippincott & Co., 1878, pp. 27 & 62.)

With such encomiums heaped upon its wares, it is small wonder that the company lost no time in capitalizing upon the awards in its advertising. A publication titled "Centennial Catalogue" issued by the company is in reality a promotional brochure illustrated by several wood-engravings, not a true catalog. It contains a brief history of the firm, a list of major commissions and, as its *pièce de résistance,* a picture and description of a chandelier installed in the Western Union Telegraph Company Building in New York City. The description reads as follows:

> We quote from Appleton's *Art Journal* of August, 1875, its description of an elegant Chandelier, kindly illustrated under the head of AMERICAN ART MANUFACTURES:
>
> "We engrave, as the second of the series of illustrations of American Art manufactures, an original design of a gas chandelier, executed by Messrs. MITCHELL, VANCE & CO., for the main entrance of the new building of the Western Union Telegraph Company, recently erected in this city. The style represents the early Greek form of ornamentation, and is highly creditable to the house by which it was produced. The main stem consists of a tapering pedestal ornamented with female figures in low relief, supporting a gracefully designed Greek vase garlanded with laurel-wreaths. — Above the top of the vase the stem is richly-ornamented and is crowned with a canopy formed by a succession of lions' heads in high-relief, holding gilt curb-rings in their jaws. Surrounding the stem are four fluted columns resting upon ornamental bases. These columns have richly-foliated capitals, and support a dome-like structure, upon which are perched a series of flying non-descript animals. Between the columns are four griffins, and from the pedestals which support them and the bases of the columns spring the several arms. The burners represent antique lamps, and are ornamented with shades in harmony with the general design. The chandelier is massive in appearance, but graceful

withal, and is finished in the style known as verd-antique, and relieved at prominent points by judicious gilding. It has eight lights, and is one of the most elaborate designs of the kind ever executed in this country. The drawings were made by Mr. Charles C. Perring, chief designer for the company."

The "non-descript" animals appear to be winged lions from whose brows spring unicorn's horns. The description of this "highly creditable" design reveals the mid-Victorian fondness for elaboration, regardless of style. Modern taste would readily grant that the fixture was "massive in appearance" but would perhaps hesitate to concede that it was "graceful withal." The Western Union Telegraph Company headquarters, in whose entrance the ponderous eight-light chandelier was installed, was designed by George B. Post and completed in 1875 at the northwest corner of Broadway and Dey Street. Along with Richard Morris Hunt's New York Tribune Building of the same year, also furnished with Mitchell, Vance & Company fixtures, it was one of the two tallest buildings in New York City, a proto-skyscraper of ten and a half stories.

Most fixtures produced by the various American manufacturers before the Civil War were in the neo-Rococo, neo-Baroque or neo-Renaissance styles or eclectic blends thereof, often with allegorical statuettes as ornaments. Charles Locke Eastlake, whose *Hints on Household Taste* came out in London in 1868 and was first published in the United States in 1872, had a tremendous influence on design during the 1870s and 1880s, but his reformist principles (often misinterpreted in America) had no apparent effect on the production of Mitchell, Vance & Company, at least not until after the Centennial year. Charles C. Perring, the firm's chief designer, clearly favored the *Néo-Grec* style that had become widespread during the 1860s in the French Second Empire. Possibly that indicates Parisian training, but nothing is known of Perring or the length of his service with Mitchell, Vance & Company. A turn-of-the-century catalog of the company's products in the Philadelphia Athenaeum illustrates much lighter and more delicate combination gas and electric fixtures fitted with gas candles.

The firm did an extensive business in bronze and brass church fittings as well as in specially designed lighting fixtures. Their 1876 brochure prefaced a list of sixty-nine clients with the following sentence:

> We do not deem it necessary to refer to any of the numerous elegant Private Dwellings in the City of New York, and throughout all the cities of the country we have supplied with Gas Fixtures, but would respectfully invite attention to the following Hotels, Public Halls, Churches, &c., (a few among very many others) as furnishing examples of styles and character of Gas Fixtures and Metal Work specially designed and manufactured by us.

The list of sixty-nine buildings included twenty-four churches, twenty hotels and ten theaters in addition to other outstanding structures such as the Western Union Building and New York Tribune Building already mentioned.

Among the New York City places of worship fitted up by Mitchell, Vance & Company were the Church of the Holy Trinity, designed by Leopold Eidlitz for the Rev. Dr. Stephen H. Tyng and completed in 1874 on the northeast corner of Madison Avenue and 42nd Street; W. Wheeler Smith's 1872 Collegiate Church of St. Nicholas on Fifth Avenue, where part of Rockefeller Center now stands; Richard Upjohn's St. Thomas's Church of 1870 on the site of the present edifice; James Renwick's masterpiece, St. Patrick's Cathedral; the 1868 Temple Emanu-El by Eidlitz and Henry Fernbach on the northeast corner of Fifth Avenue and 43rd Street; and Upjohn's Trinity Chapel of 1855 on West 26th Street. H. H. Richardson's Brattle Square Church of 1873, now the First Baptist Church at Commonwealth Avenue and Clarendon Street in Boston, also appears among the architecturally significant churches on that ecumenical list.

The twenty hotels included such then fashionable but now vanished hostelries as the Buckingham, Windsor, Westminster and St. Denis, as well as the Astor House, Grand Central and Gilsey House among the twelve New York hotels listed. The vast United States Hotel at Saratoga Springs, Chicago's Palmer House and Grand Pacific Hotel and the Galt House in Louisville were nationally known in their day and represented the cream of hotel patronage for Mitchell, Vance & Company. The ten theaters listed included six in New York City, among them the theater completed for Edwin Booth in 1869 at Sixth Avenue and 23rd Street and the Grand Opera House of 1868, built for Samuel Pike at Eighth Avenue and 23rd Street. Whitney's Opera House in Detroit and the theater built for "Lucky" Baldwin in San Francisco also appear on the list.

Leading New York retail establishments for which the firm designed and manufactured gas fixtures included Wheeler & Wilson's sewing-machine emporium on Union Square and Lord & Taylor's splendid new store of 1873 at Broadway and 20th Street. (Lord & Taylor's previous store of 1859 at Broadway and Grand Street had featured a huge gaselier [i.e., a gas chandelier] made by Tiffany's to light the staircase.) Bryant and Gilman's Boston City Hall of 1862, and the Equitable Life Assurance Company building by N. J. Bradlee, completed in Boston in 1874, as well as the New Illinois State Capitol in Springfield, "Smith's Female College" in Northampton, Massachusetts and Harvard's Memorial Hall of 1874–1876 by Ware and Van Brunt were other major American buildings lighted by Mitchell, Vance & Company's fixtures. The great corona designed for Sanders Theater in Memorial Hall at Harvard still remains in place.

Before leaving the 1876 brochure that was somewhat inaccurately titled "Centennial Catalogue," it should be noted that its text provides clues to a number of contemporary presuppositions regarding the appropriate application of design to gas fixtures, e.g.: "For the Reception Room, Chandeliers in Gold, relieved with a little color—as jet, crimson, or blue are deemed desirable." A matching twelve-light fixture was suggested for the drawing room, and chandeliers with center slides were recommended for "other rooms," presumably the library and dining room, since center slides permitted light to be lowered close to center tables. A center slide usually contained an Argand burner, while the branches of the same fixture supported "fishtail" or "batswing" burners (so-called from the shapes of their open flames). The *Néo-Grec* style predominated among the illustrations in the publication. A *Néo-Grec* "slide Library Chandelier" was ornamented with medallions representing music, poetry and history, but the substitution of medallions representing game, birds and fish could render it suitable for the dining room. That particular example was available in bronze, gilt or verde antique finish.

At least ten catalogs are known to have been issued by Mitchell, Vance & Company, since the cover of the undated turn-of-the-century example in the Philadelphia Athenaeum reads, "Combination Fixtures Catalogue No. 10." It contains only thirteen plates, numbered 211 through 224. The title page reads as follows:

> The Mitchell Vance Company, Manufacturers of every Description of Gas and Electric Light Fixtures, Etc. Dealers in All goods appertaining to the Gas and Electric Fixture Trade. Manufactory: 10th Avenue, 24th & 25th Sts. Salesrooms: 836 & 838 Braodway and 13th St. New York.

Many of the photographic illustrations show combination fixtures with "gas candles" formed by burner tips at the ends of pipe lengths masked by white porcelain candle-shaped sleeves.

One of the most usual forms of catalog issued between the 1850s and the 1890s by American manufacturers of gas fixtures consisted of a series of lithographed plates, originally unbound. The system had the advantage of being readily updated by the addition of new plates and the withdrawal of obsolete ones without having to go to the trouble of issuing an entirely new book at stated intervals. There was also the advantage of flexibility, since jobbers or retail customers needed only the plates pertaining to their specific needs instead of an entire book.

There are two Mitchell, Vance & Company catalogs at the great museum-library of American arts in Winterthur, Delaware, one with thirty-one plates numbered 540 through 571, and another with thirty plates showing "crystal glass chandeliers, brackets, standards, etc." Thus, the series of eighty-three lithographs (bound together in this case) reproduced in the following pages comprises the most extensive collection of Mitchell, Vance & Company ephemera that has so far been discovered. It is also certainly the most typologically varied, as every kind of gas appliance and lighting fixture made before the 1880s, except crystal fixtures, is represented. All but sixteen of the original plates bear the name of the firm that printed them, variously given as Brett Litho. Co. (the most frequent in occurrence, with 37 examples), Brett & Fairchild Lith. and four other minor variations; the address is usually 116 Fulton Street, but in six cases it is 83 Nassau Street. The changing names and addresses suggest the fluidity of nineteenth-century commercial charters, as well as indicating that the plates bound in the volume were produced over a number of years.*

The first plate in this group of Mitchell, Vance & Company presentations was obviously intended for gas fitters rather than for householders. The same also applies to Plate 123 (p. 39 of this edition). Note that Plate 1 (p. 1) includes a street-lamp cock. The specialization of the various types of gas cocks indicates a highly sophisticated standard of manufacture. Among the surprises to be found in the plates is the group of gas stoves and "Ætnas" shown in Plates 36 and 37 (pp. 31 and 32). Gas stoves were just beginning to "come in" during the 1870s, as the patent date of July 14, 1868 on the oven suggests.

During the 1860s and 1870s, the I. P. Frink Company in New York had the lion's share of the reflector fixture trade, but Plates 296, 297 and 298 (pp. 81, 82 and 83) show that Mitchell, Vance & Company was among their competitors. Note the specialized design of the show-window reflector fixtures on Plate 296 (p. 81). The "Grand Prismatic Reflecting Sunlight" with corona extra on the same plate must surely have been at the top of the firm's reflector line. Ornamental reflector street lamps and splendid store signs were also available, as shown on Plate 298 (p. 83). Elegant gas lanterns in square, hexagonal, octagonal, circular and globular forms for street lights in front of restaurants, theaters, hotels, stores and other places of public assembly are shown on Plates 28 and 28½ (pp. 23 and 24). Note that the restaurant lantern on Plate 28 (p. 23) is a model named "The Prince Albert."

Plates 27, 27½, 193 and 293 (pp. 21, 22, 45 and 78) provide evidence that Mitchell, Vance & Company manufactured glass shades, or globes, in addition to the "crystal glass" fixtures for which they received an accolade from the Centennial Commission judges. As the firm doubtless supplied jobbers and retailers with shades in bulk, the presence of a Mitchell, Vance & Company shade on a fixture is insufficient evidence for a secure attribution of the fixture itself unless other corroborating evidence exists. Shades of a different sort are shown on Plate 22 (p. 16). The metal frames (numbers 1 and 2) for "Porcelain Plates" almost certainly held lithophanes (porcelain with figures made distinct by light passing through them). Note that the metal pulpit globe and shade were designed to concentrate light on a manuscript. Except for "portables," or gas lamps, and "billiards" to light billiard tables, almost all gas jets were sheltered from drafts by shades. "Portables" are shown on various plates mentioned below, and "billiards," which had horizontal instead of vertical jets, are shown on Plates 19½, 21 and 279 (pp. 13, 15 and 65), as well as on Plate 22.

Glass smoke bells, used primarily for hall fixtures and occasionally to protect painted ceilings, are illustrated on Plate 27 (p. 21). A porcelain example appears on Plate 295 (p. 80). The patterns of number 318 on Plate 27½ (p. 22), 110 and 112 on Plate 193 (p. 45) and 426, 6811½ and 6796 on Plate 293 (p. 78) are decidedly *Néo-Grec* in character. Metal coronets like that on Plate 193 (available in either gilt or bronze finish) were occasionally used with very elaborate fixtures like those shown on Plates 170 and 171 (pp. 40 and 41), but they rarely appear in domestic settings; their use seems to have been mostly restricted to public buildings. It should be noted that *all* of the shades have bases of small diameter. Wide-throated shades were not made until after 1875, but by the early 1880s they had almost entirely supplanted the small-based type, since the wider diameter of their bases greatly reduced flickering by providing an even flow of air to the burner tips.

Glass shades were normally ornamented either by frosting and etching done with the aid of stencil patterns, or by cutting, or by a combination of both techniques. Etching could be deep or shallow, depending on how long the acid was allowed to react with the glass. Bolder patterns, such as that of number 418 on Plate 293 (p. 78), were probably cut, and delicate patterns like number 6809½ on the same plate were usually etched. Occasionally shades were tinted or hand-painted. Those described as decorated were certainly painted. The porcelain shades on Plate 295 (p. 80) were, except for the last two before the smoke bell, intended for use on the central lights of center-slide fixtures like those on Plates 17, 18½, 264, 282 and 291 (pp. 10, 11, 50, 68 and 77). The extension lights of center-slide fixtures had Argand burners, which required the conical

* To allow more room for the illustrations, the present edition omits from the plate pages the name and address of the printer as well as the name and address of Mitchell, Vance (597 Broadway). Each Dover page bears, in new type, the Dover page number followed in parentheses by the original plate number—for example, 27 (Plate 32). With regard to the original plate headings, see the Note following this Introduction.

form of shade. The hexagonal shade, number 851, on Plate 295, was for a lamp and contained lithophanes.

Other glass and porcelain fittings, bobêches, prisms and a "candle" of the kind used to slip over an elongated burner pipe, are shown on Plate 42 (p. 37). That plate presents a miscellany of things, including a "Kero Cup" holder and a fount for equipping a gas fixture to support oil lamps, "cloth brackets," probably intended to hold towels, and chandeliers ranging in design from the relatively simple number 1041 to the quite elaborate number 2470. Curiously, Plates 3 and 120 (pp. 3 and 38) present almost precisely the same subject matter. Both show the same pendants, as simple two-light ceiling fixtures were called, three drop lights to bring light from chandeliers closer to desks or tables, a one-light extension or slide pendant at the extreme left, two hall lights (numbers 757 and 759) and several lighters with tubes holding tapers and bifurcated flanges to turn gas keys. The decoration varies between the two plates in many details, however.

A wide selection of hall fixtures, or "hall pendants," is shown on Plates 19, 19½ and 20 (pp. 12, 13 and 14). A "billiard" (light) also appears on Plate 19½, and there are two "store pendants" on Plate 20. Other hall pendants appear at random throughout this series of plates, two of the most elaborate appearing on Plate 38 (p. 33). Two-light pendants, often specifically designated as "store pendants," also appear at random, ranging in type from the very simple examples shown on Plates 4, 21, 29 and 275 (pp. 4, 15, 25 and 61) to relatively elaborate ones like those on Plates 5½, 6, 6½, 31, 32, 277 and 286 (pp. 5, 6, 7, 26, 27, 63 and 72). Another type of pendant was the "toilet," a chandelier, usually with three burners, suspended from a bracket cantilevered out from the wall. "Toilets" were designed to be suspended near dressing tables to light their mirrors without obstructing them. Six examples appear on Plate 274 (p. 60), two on Plate 273, (p. 59), two on Plate 291 (p. 79) and one each on Plates 34 and 287 (pp. 29 and 73).

Wall brackets, never called "sconces" during the gas age, were quite as common as pendants and other ceiling fixtures. Plates 2, 23, 24 and 288 (pp. 2, 17, 18 and 74) are devoted entirely to brackets. Most single brackets were "swing brackets," that is, they could be moved from side to side, and many of the plainest were two- or three-jointed fixtures that could be extended horizontally. Numerous examples of jointed brackets appear on Plate 2. They were frequently used to bring lights close to mirrors in bedrooms, or close to desks in offices, and they were also much used in the service areas of dwellings. The word "stiff" applied to numbers 798, 1142 and 104½ on Plate 2 and numbers 1650, 1652, 1654 and 1662 on Plate 288 indicates immovable brackets. Two-light brackets, examples of which are shown on Plate 24, were always "stiff." Note that all of those two-light brackets could "be furnished in 3 Lights." Normally, the more elaborate brackets were made to go *en suite* with chandeliers. Five brackets of unusual design, three with match holders termed "match boxes" and two incorporating human arms and hands in their pattern, appear on Plate 18½ (p. 11).

Plate 17 (p. 10) shows extension chandeliers with a patent jointed pipe and spring-reel device. Such fixtures, like center-slide chandeliers, were intended for use over center tables or flat-topped desks. Two center-slide chandeliers with the spring-reel device are shown on Plate 18½ (p. 11). Plate 264 (p. 50) shows three counterweighted center-slide chandeliers, and a rare single-light counterweighted slide pendant (number (1147) appears on Plate 279 (p. 65). Two counterweighted pendants (numbers 2792 and 2794) and a spring-reel pendant (number 2790) are shown on Plate 290 (p. 76). Most of the various types of vertically adjustable fixture used telescoping pipe instead of the complex jointed pipes used with the spring-reel fixtures. A few fixtures were adjustable in almost any direction through the use of universal joints, as in number 748 on Plate 7½ (p. 8), an example with an Argand burner.

The most flexible of all gas lights was the "portable," or gas lamp, which was attached to a fixture by a rubber hose. Some portables exhibited the most unrestrained fantasy of which designers were capable. Plate 7½ (p. 8) shows one portable in the form of a sphinx, a favorite motif of the period, and another with the figure of an infant, although neither quite attains the surrealistic effect of two brackets on the same plate, one composed of a hand holding a Roman lamp (number 1340) and the other (number 1446) representing a chained dog who appears preoccupied with whatever is on his mind. A very similar dog, now breathing fire, appears as a cigar lighter (number 01272) on Plate 289 (p. 75). Plate 25 (p. 19) is devoted to gas lamps ("portables"), both plain and with statuettes. Among the latter are representations of a Minuteman (number 0838), a "Young Patriot" (number 0242) and Benjamin Franklin (number 0108), among others. The portables at the top of Plate 31 (p. 26) include an angel, perhaps intended for a nursery. Numbers 01066 and 01064 on Plate 34 (p. 29) may possibly represent Oberon and Titania. (The angel bracket on the same plate supports a cluster of gas candles of the type shown also on Plate 280 [p. 66].) The four portables shown on Plate 39 (p. 34) are particularly fine examples of their statuette type. Number 01108 on Plate 40 (p. 35), a "sliding portable," has the familiar student-lamp form.

Cigar lighters were at least as fanciful as lamps in design. Plate 26 (p. 20) shows a number of cigar lighters, among them examples featuring a Chinese (number 0534), a young sailor (number 0970), a vivandière (number 0536), Mercury (number 0962) and a fire hydrant (number 096), perhaps intended for the volunteer fire-company trade, as well as a stag, a dog and a horse. Note that the three cigar lighters on Plate 40 (p. 35) were fitted to burn alcohol instead of gas.

The various types of ceiling and wall fixture have all been mentioned, but one type of lighting fixture, the "standard," remains to be discussed. Standards were affixed to counters, pulpits, desks, railings and newel posts and placed in niches. They ranged in design from the relatively simple examples shown on the lower half of Plate 26 (p. 20) to such splendid ones as number 0924 on Plate 170 (p. 40), the three on Plate 171 (p. 41), and the lavish standard numbered 01256 on Plate 284 (p. 70). Note that the two standards on Plate 39 (p. 34) have cigar-lighter attachments, indicating that they stood upon counters or, perhaps, hotel registration desks.

A special form of standard was the figural type illustrated on Plates 216, 217 and 254 (pp. 46, 47 and 48), captioned "Figures for Gas." Most supported a single burner, but the pair of Indians on Plate 271 (p. 57) could support up to five lights apiece. The pair of figures on Plate 267 (p. 53) appears, because of the dead-game motif, to have been intended for a dining room, perhaps on a built-in sideboard or in wall niches. The griffin standard on Plate 269 (p. 55) provides a change from the more usual figures of human beings, and the bracket representing Mercury on that plate is the same in design concept as the figural standards. These standards were almost invariably intended for use on newel posts or in stairway hall niches. The fact that so many are designed in pairs indicates that many were to be used on very grand stairways with freestanding lower runs requiring two newel posts. The three plates captioned "Figures for Gas" present figures of a warrior about to sound his horn (Roland at Roncesvalles?), a vestal guarding her fire, gladiators (Plate 216 [p. 46]), Renaissance courtiers, goddesses, knights (Plate 217 [p. 47]) and rustic youths and cavaliers (Plate 254 [p. 48]). The choice of subject matter suggests that the romanticism of the previous generation was by no means dead more than a decade after the end of the Civil War. Certainly sentimentality was flourishing.

The vogue for ornamenting chandeliers with statuettes that was so prevalent in the 1850s lingered on with greatly reduced intensity into the 1870s. Vestiges of it remained in such examples as the birds perched on chandelier number 2664 on Plate 268 (p. 54), the figure within the arched frame of the chandelier on Plate 271 (p. 57), the statue of Diana, goddess of the hunt, in the chandelier on Plate 272 (p. 58) and the sphinxes atop the branches of the pair of chandeliers on Plate 270 (p. 56). The chandelier with the birds was probably intended for a parlor or boudoir, and the chandelier with Diana, which is also ornamented with stag's heads, was certainly designed for a dining room. The matched pair with the sphinxes on Plate 270 (p. 56) would have been used in a pair of rooms *en suite,* either a parlor and library, or a parlor and dining room, since one of the two chandeliers has a counterweighted center-slide Argand light to be lowered over a table.

As noted previously, ceiling fixtures with one or two burners were termed "pendants." Hall pendants usually had one burner, and store pendants normally had two. "Billiards" and reflectors were special categories among ceiling fixtures. All other ceiling fixtures with three or more burners were called chandeliers. Earlier they had been called gaseliers, also spelled gasolier or gasalier, but by 1876 that term was obsolete. Chandeliers most frequently had three, four or six burners. Except in extraordinarily large houses, twelve lights was about the limit for chandeliers in domestic use. Particularly attractive examples of six-light chandeliers may be seen on Plates 35 and 36 (pp. 30 and 31). Twelve-light fixtures were two-tiered. A decidedly "Eastlake" example appears in the center of Plate 16 (p. 9), and others of more *Néo-Grec* character are shown on Plates 263 and 283 (pp. 49 and 69). Plate 181 (p. 43) is entirely given to one especially grand two-tiered twelve-light chandelier.

The fixtures on Plates 170 and 171 (pp. 40 and 41) were apparently designed for public buldings, as few private houses required five-light brackets like number 1084 or standards like number 0920. Those fixtures and the large chandelier on Plate 179 (p. 42) are all consistent in their Renaissance Revival style, although they do not actually match in design. The chandeliers on Plates 179, 192 and 284 (pp. 42, 44 and 70) were certainly intended for public places, as they could carry as many as thirty-two, sixty-four and forty lights respectively. Those singularly splendid, if somewhat ponderous, fixtures must have appeared grand and impressive indeed in the interiors they once illuminated.

Gas fixtures of the more elaborate sort were not monochromatic. Even if they were entirely gilt, the gilding was usually partly matte and partly burnished. The variety of colors available included blue, "verde," "verde antique," "light verde," maroon, black and pink, generally combined with gilt or bronze. A plate of color samples (not reproduced here) cautions that "To produce the best effect by these colors, it is important that the design be of an appropriate character." The rival firms of the period produced similar combinations of finish.

Until the close of the last century, gas lighting was the most reliable and popular artificial illumination available. After Thomas Alva Edison produced a practical incandescent electric bulb late in 1879 and established the first central power station in New York City in 1882, electric lighting made slow progress against gas, because the Welsbach burner, or gas mantle, first manufactured commercially in 1887, soon made incandescent gas lighting competitive with the new electric lighting. The flat-flame fishtail and batswing burners (so-called from the shapes of their open flames) that had been previously used, and that appear throughout this catalog, were rendered obsolete not only by electric bulbs but also by Welsbach's mantles, which permitted gas to retain its lead as an illuminant into the present century, when Mitchell, Vance & Company turned to making combination gas and electric fixtures and, eventually, electric fixtures only, before going out of business in 1933.

DENYS PETER MYERS
Consulting Architectural Historian

Note

To allow more room for the illustrations, the present edition omits the original plate headings. In the majority of cases, this heading was merely "Gas Fixtures." Following is an alphabetical list of other original headings, together with the *Dover page numbers* that correspond to those headings:

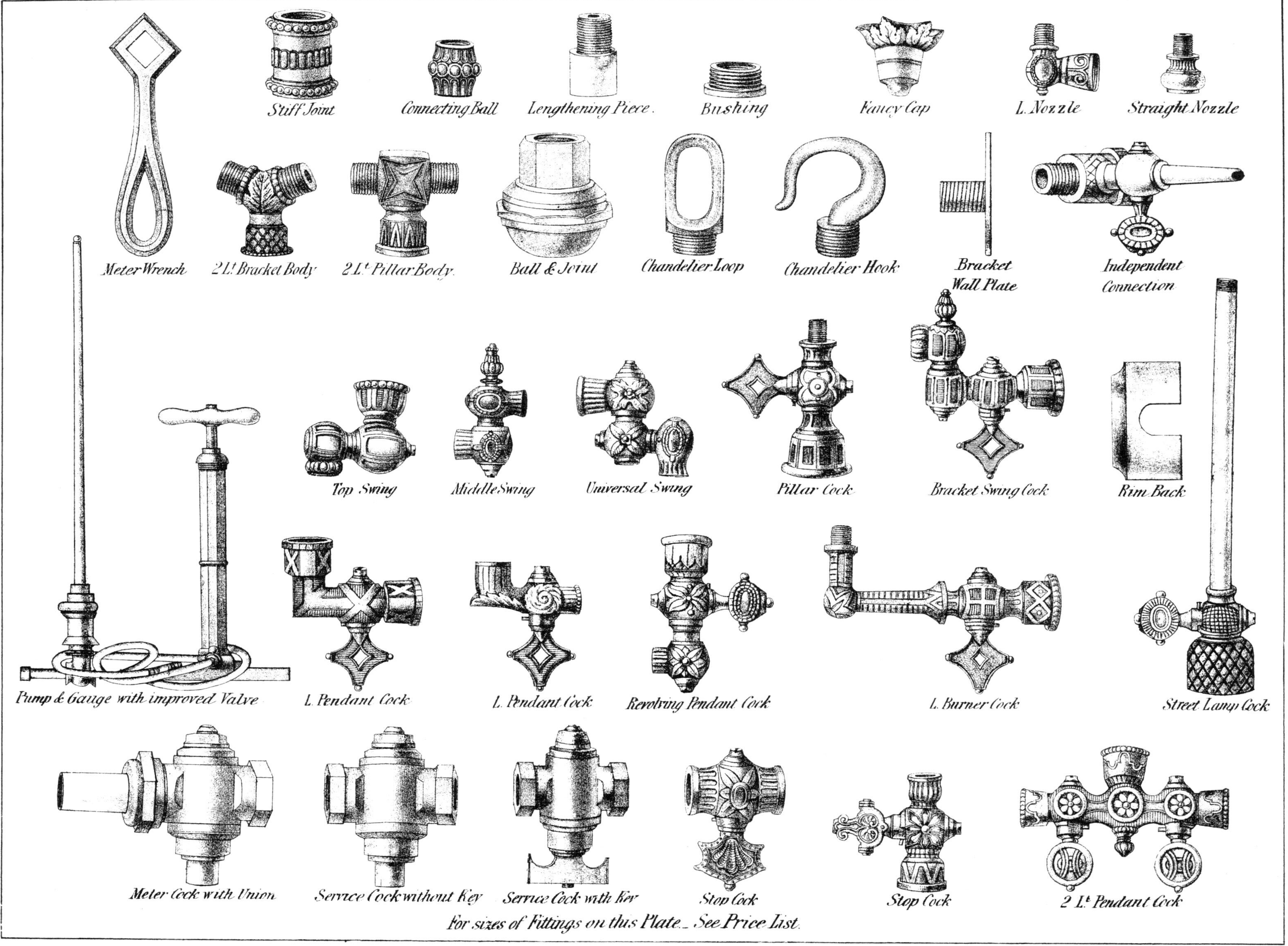

For sizes of Fittings on this Plate _ See Price List.

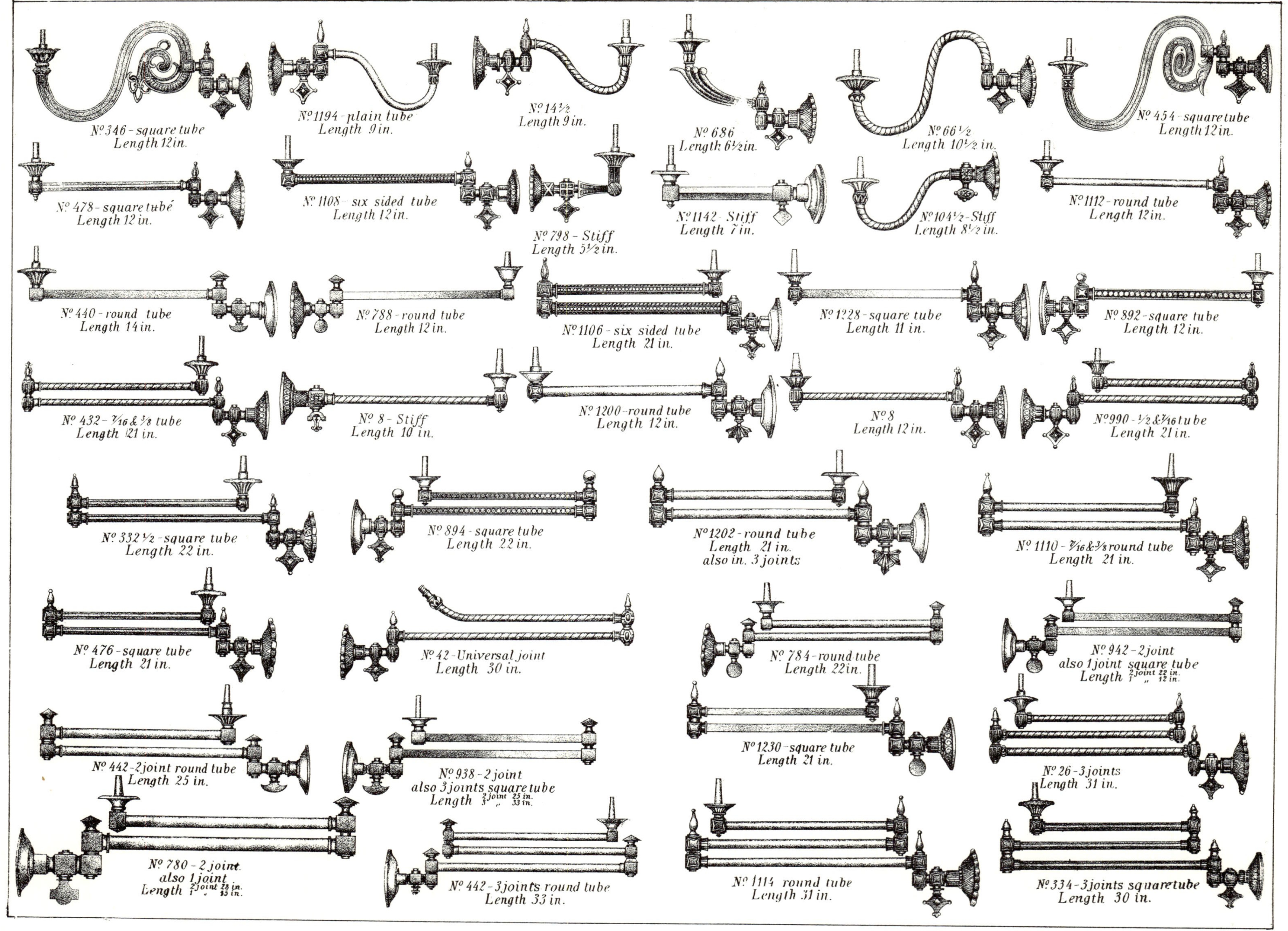
No. 346 - square tube
Length 12 in.
No. 1194 - plain tube
Length 9 in.
No. 14½
Length 9 in.
No. 686
Length 6½ in.
No. 66½
Length 10½ in.
No. 454 - square tube
Length 12 in.
No. 478 - square tube
Length 12 in.
No. 1108 - six sided tube
Length 12 in.
No. 798 - Stiff
Length 5½ in.
No. 1142 - Stiff
Length 7 in.
No. 104½ - Stiff
Length 8½ in.
No. 1112 - round tube
Length 12 in.
No. 440 - round tube
Length 14 in.
No. 788 - round tube
Length 12 in.
No. 1106 - six sided tube
Length 21 in.
No. 1228 - square tube
Length 11 in.
No. 892 - square tube
Length 12 in.
No. 432 - 7/16 & 3/8 tube
Length 21 in.
No. 8 - Stiff
Length 10 in.
No. 1200 - round tube
Length 12 in.
No. 8
Length 12 in.
No. 990 - 1/2 & 7/16 tube
Length 21 in.
No. 332½ - square tube
Length 22 in.
No. 894 - square tube
Length 22 in.
No. 1202 - round tube
Length 21 in.
also in. 3 joints
No. 1110 - 7/16 & 3/8 round tube
Length 21 in.
No. 476 - square tube
Length 21 in.
No. 42 - Universal joint
Length 30 in.
No. 784 - round tube
Length 22 in.
No. 942 - 2 joint
also 1 joint square tube
Length 2 joint 22 in. 1 " 12 in.
No. 442 - 2 joint round tube
Length 25 in.
No. 938 - 2 joint
also 3 joints square tube
Length 2 joint 25 in. 3 " 33 in.
No. 1230 - square tube
Length 21 in.
No. 26 - 3 joints
Length 31 in.
No. 780 - 2 joint
also 1 joint
Length 2 joint 28 in. 1 " 15 in.
No. 442 - 3 joints round tube
Length 33 in.
No. 1114 round tube
Length 31 in.
No. 334 - 3 joints square tube
Length 30 in.

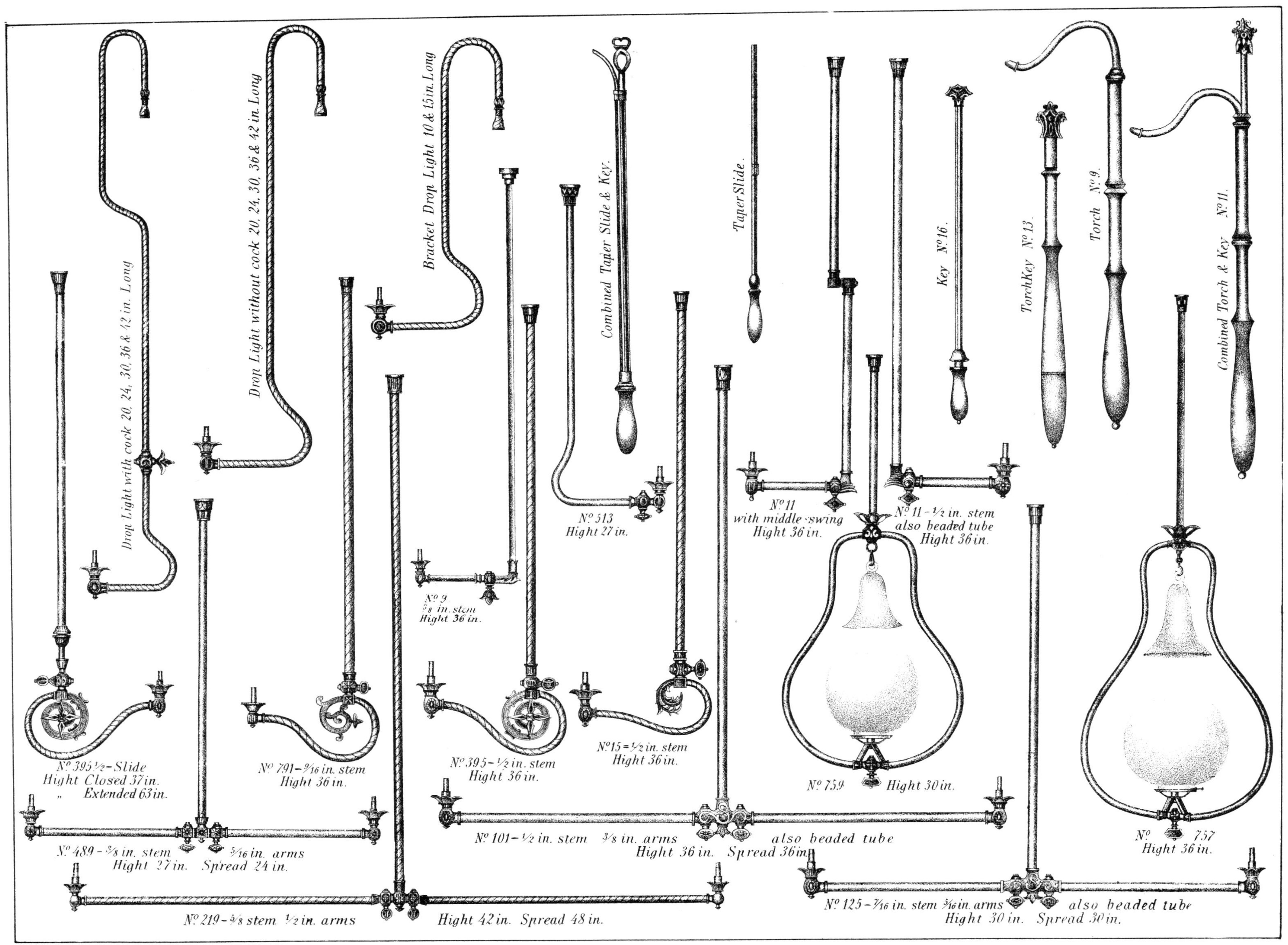
Drop Light with cock 20, 24, 30, 36 & 42 in. Long
Drop Light without cock 20, 24, 30, 36 & 42 in. Long
Bracket Drop Light 10 & 15 in. Long
Combined Taper Slide & Key.
Taper Slide.
Key No. 16.
Torch Key No. 13.
Torch No. 9.
Combined Torch & Key No. 11.
No. 513
Hight 27 in.
No. 11
with middle swing
Hight 36 in.
No. 11 - 1/2 in. stem
also beaded tube
Hight 36 in.
No. 9
3/8 in. stem
Hight 36 in.
No. 15 = 1/2 in. stem
Hight 36 in.
No. 395 1/2 - Slide
Hight Closed 37 in.
" Extended 63 in.
No. 791 - 9/16 in. stem
Hight 36 in.
No. 395 - 1/2 in. stem
Hight 36 in.
No. 759 Hight 30 in.
No. 757
Hight 36 in.
No. 489 - 3/8 in. stem 5/16 in. arms
Hight 27 in. Spread 24 in.
No. 101 - 1/2 in. stem 3/8 in. arms also beaded tube
Hight 36 in. Spread 36 in.
No. 219 - 5/8 stem 1/2 in. arms Hight 42 in. Spread 48 in.
No. 125 - 7/16 in. stem 5/16 in. arms also beaded tube
Hight 30 in. Spread 30 in.

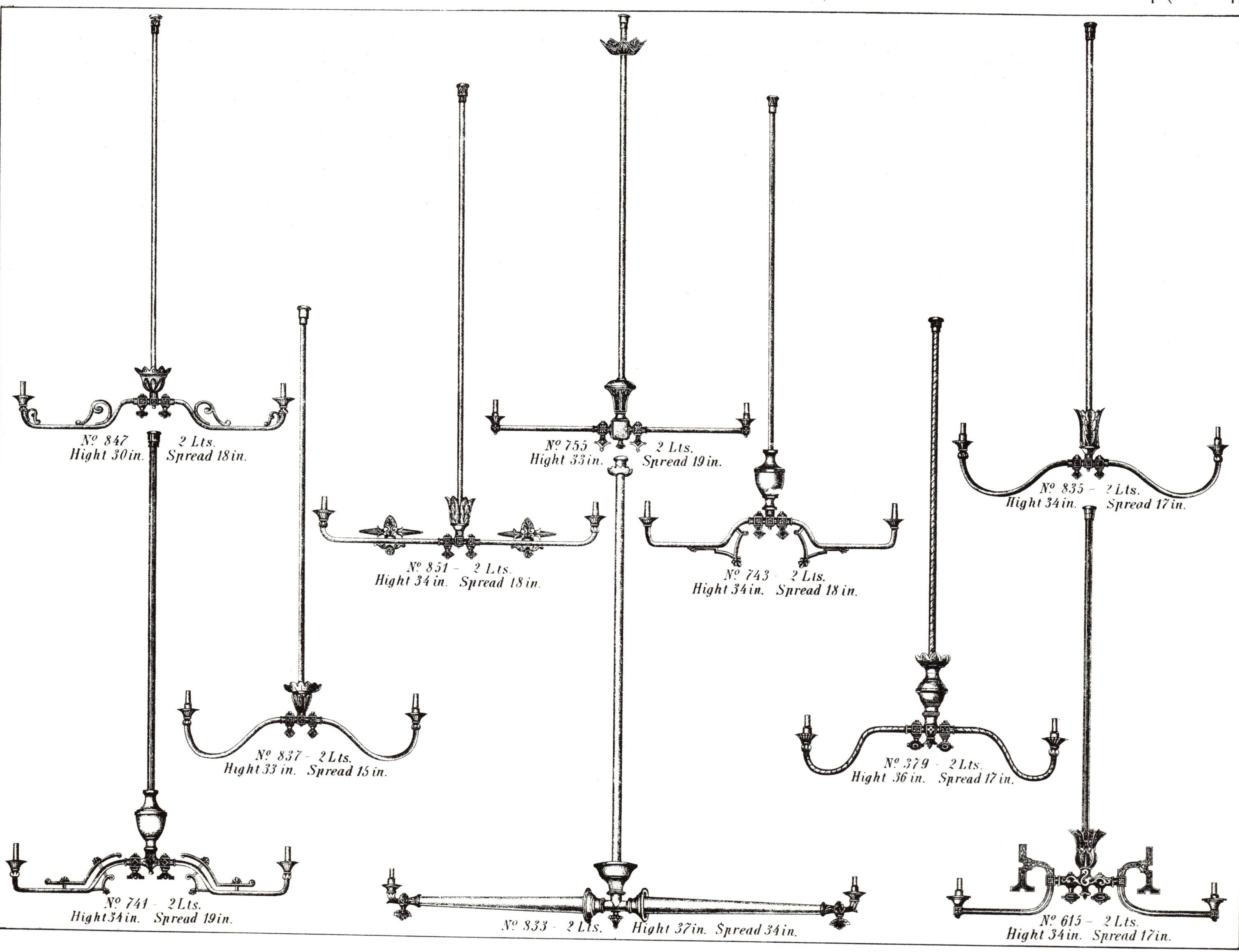
No. 847 2 Lts.
Hight 30 in. Spread 18 in.
No. 755 2 Lts.
Hight 33 in. Spread 19 in.
No. 835 - 2 Lts.
Hight 34 in. Spread 17 in.
No. 851 - 2 Lts.
Hight 34 in. Spread 18 in.
No. 743 - 2 Lts.
Hight 34 in. Spread 18 in.
No. 837 - 2 Lts.
Hight 33 in. Spread 15 in.
No. 379 - 2 Lts.
Hight 36 in. Spread 17 in.
No. 741 - 2 Lts.
Hight 34 in. Spread 19 in.
No. 833 - 2 Lts. Hight 37 in. Spread 34 in.
No. 615 - 2 Lts.
Hight 34 in. Spread 17 in.

Nº 1406-1 Lt.
stiff. Length 6 in.
Nº 1448-1 Lt. Length 11½ in.
Nº 1350-1 Lt.
Length 11 in.
Nº 1416-1 Lt.
stiff. Length 6 in.
Nº 2428- 2 Lt.
Hight 34 in. Spread 20 in.
Nº 2544 - 2 Lt.
Hight 34 in. Spread 15 in.
Nº 1065-2 Lt.
Hight 35 in. Spread 18 in.
Nº 2430 - 2 Lt.
also in 3 & 4 Lts.
Hight 40 in. Spread 20 in.
Nº 2484 2 Lt.
also in 3 & 4 Lts.
Hight 31 in. Spread 22 in.
Nº 2486-2 Lt. also in 3 & 4 Lts.
Hight 32 in. Spread 20 in.
Nº 2528-2 Lt. also in 3 & 4 Lts.
Hight 30 in. Spread 20 in.
Nº 2488-2 Lt. also in 3 & 4 Lts.
Hight 33 in. Spread 19 in.

2 L^ts. № 1802.
also in 3 & 4 Lts.
Hight 36 in Spread 18 in.
2 L^ts № 2158.
also in 3 & 4 L^ts
Hight 32 in. Spread 18 in.
2 L^ts № 2156.
also in 3 & 4 L^ts
Hight 32 in., Spread 18 in.
1 L^t № 865.
Hight 36 in.
Spread 12 in.
1 L^t № 879
Hight 36 in.
Spread 15 in.
2 L^ts. № 1988
also in 3 & 4 Lts.
Hight 34 in., Spread 19 in.
2 L^ts № 2250
also in 3 & 4 L^ts
Hight 33 in., Spread 18 in.
2 L^ts № 1892
also in 3 & 4 L^ts
Hight 36 in., Spread 20 in.

Nº 2498-2 Lt. also in 3 & 4 Lts.
Hight 33 in. Spread 18 in.

Nº 1418. Length 22 in.

Nº 2504 2 Lt. also in 3 & 4 Lts
Hight 30 in. Spread 17 in.

Nº 2500-2 Lt also in 3 & 4 Lts.
Hight 36 in Spread 19½ in.

Nº 2502-2 Lt also in 3 Lts
Hight 32 in. Spread 21 in.

Nº 2506-2 Lt. also in 3 & 4 Lts
Hight 30 in. Spread 17 in.

Nº 2508 2 Lt. also in 3 & 4 Lts
Hight 36 in Spread 23½ in

Nº 2510 2 Lt.
Hight 30 in. Spread 19 in.

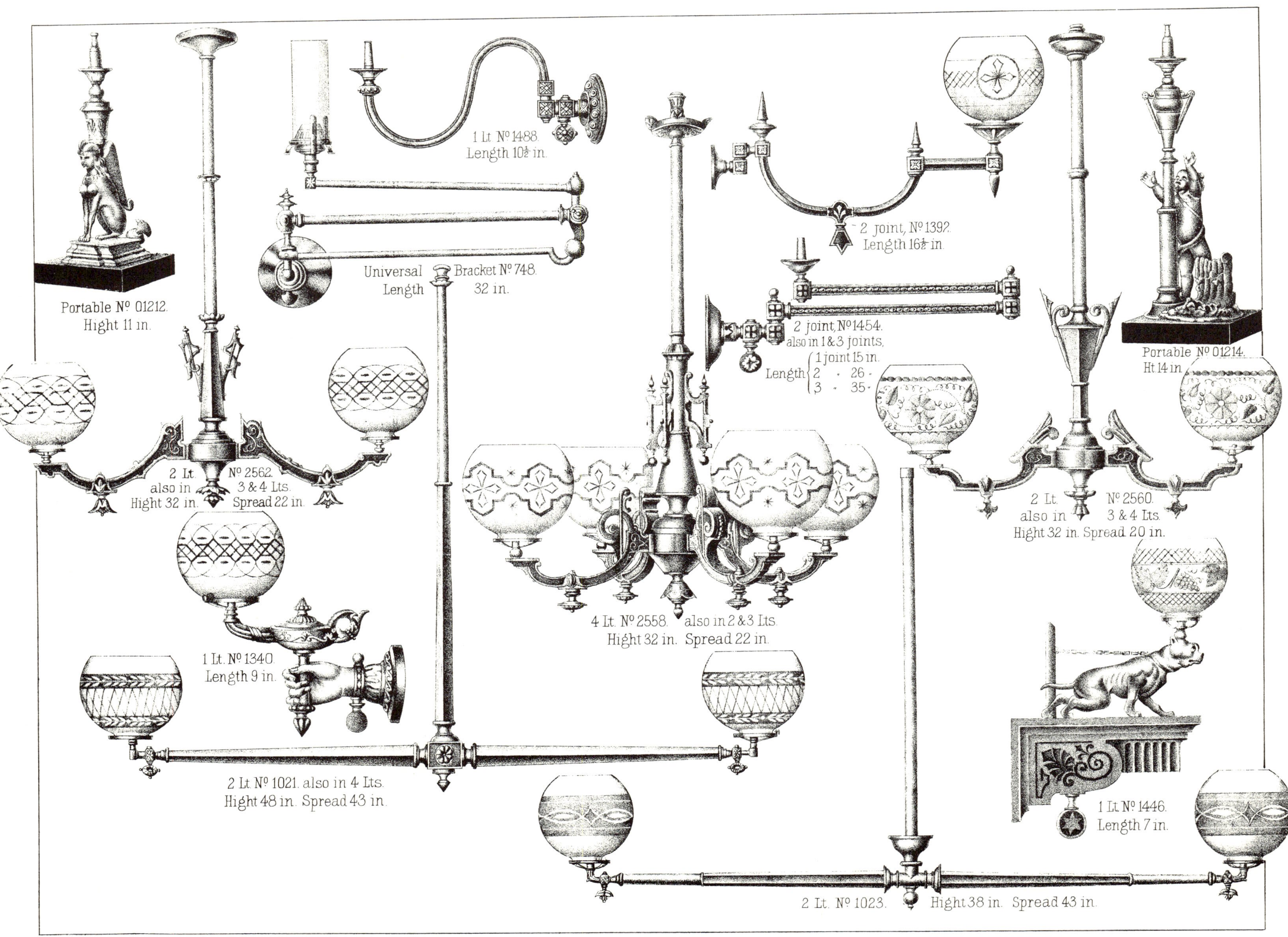

Portable Nº 01212.
Hight 11 in.
1 Lt. Nº 1488.
Length 10½ in.
Universal Bracket Nº 748.
Length 32 in.
2 joint, Nº 1392.
Length 16½ in.
2 joint, Nº 1454.
also in 1 & 3 joints,
Length 1 joint 15 in.
2 - 26 -
3 - 35 -
Portable Nº 01214.
Ht 14 in.
2 Lt. Nº 2562.
also in 3 & 4 Lts.
Hight 32 in. Spread 22 in.
2 Lt. Nº 2560.
also in 3 & 4 Lts.
Hight 32 in. Spread 20 in.
4 Lt. Nº 2558. also in 2 & 3 Lts.
Hight 32 in. Spread 22 in.
1 Lt. Nº 1340.
Length 9 in.
2 Lt. Nº 1021. also in 4 Lts.
Hight 48 in. Spread 43 in.
1 Lt. Nº 1446.
Length 7 in.
2 Lt. Nº 1023. Hight 38 in. Spread 43 in.

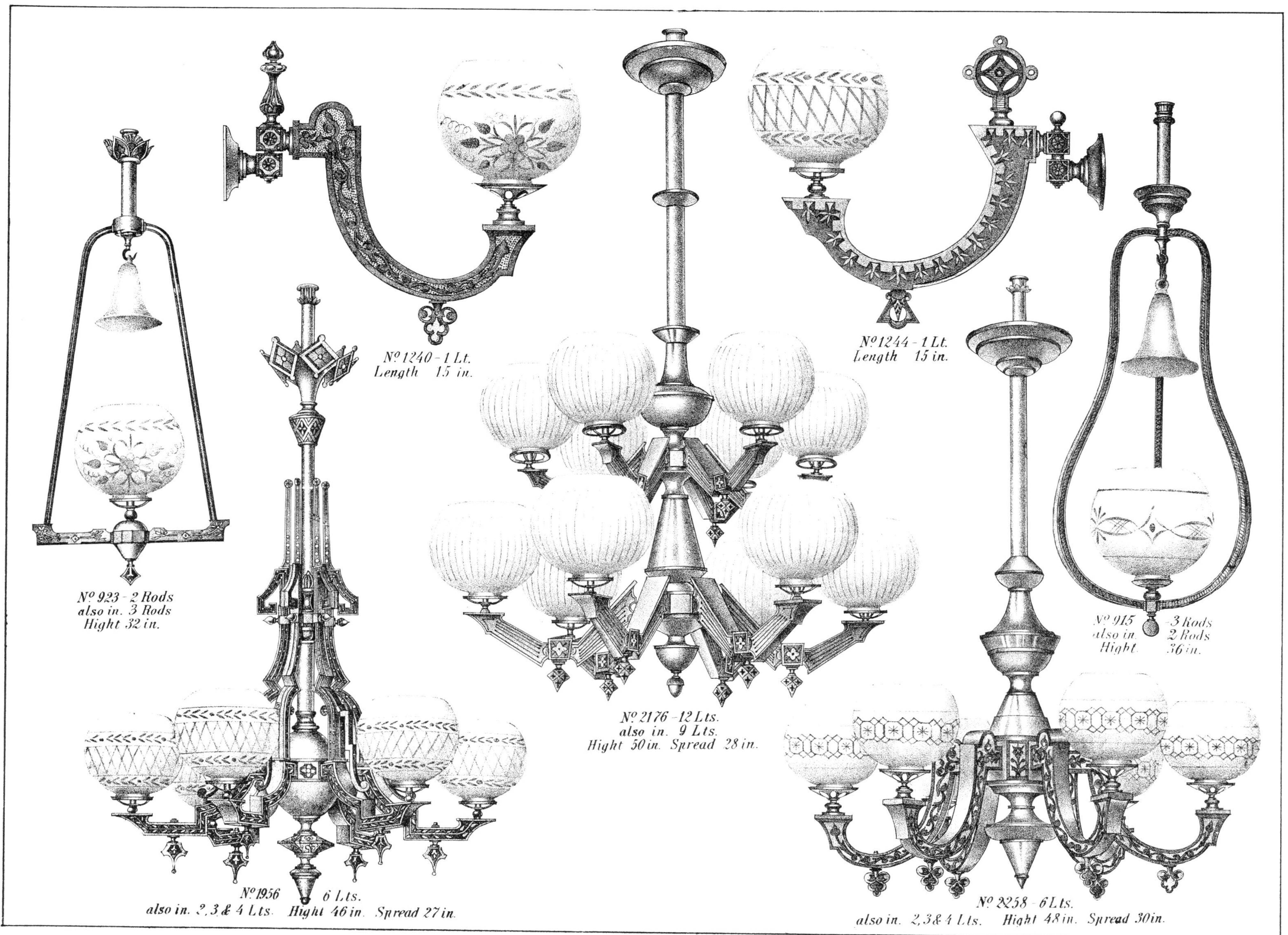

Nº 923 - 2 Rods
also in. 3 Rods
Hight 32 in.
Nº 1240 - 1 Lt.
Length 15 in.
Nº 1956 6 Lts.
also in. 2, 3 & 4 Lts. Hight 46 in. Spread 27 in.
Nº 2176 - 12 Lts.
also in. 9 Lts.
Hight 50 in. Spread 28 in.
Nº 1244 - 1 Lt.
Length 15 in.
Nº 915 - 3 Rods
also in. 2 Rods
Hight 36 in.
Nº 2258 - 6 Lts.
also in. 2, 3 & 4 Lts. Hight 48 in. Spread 30 in.

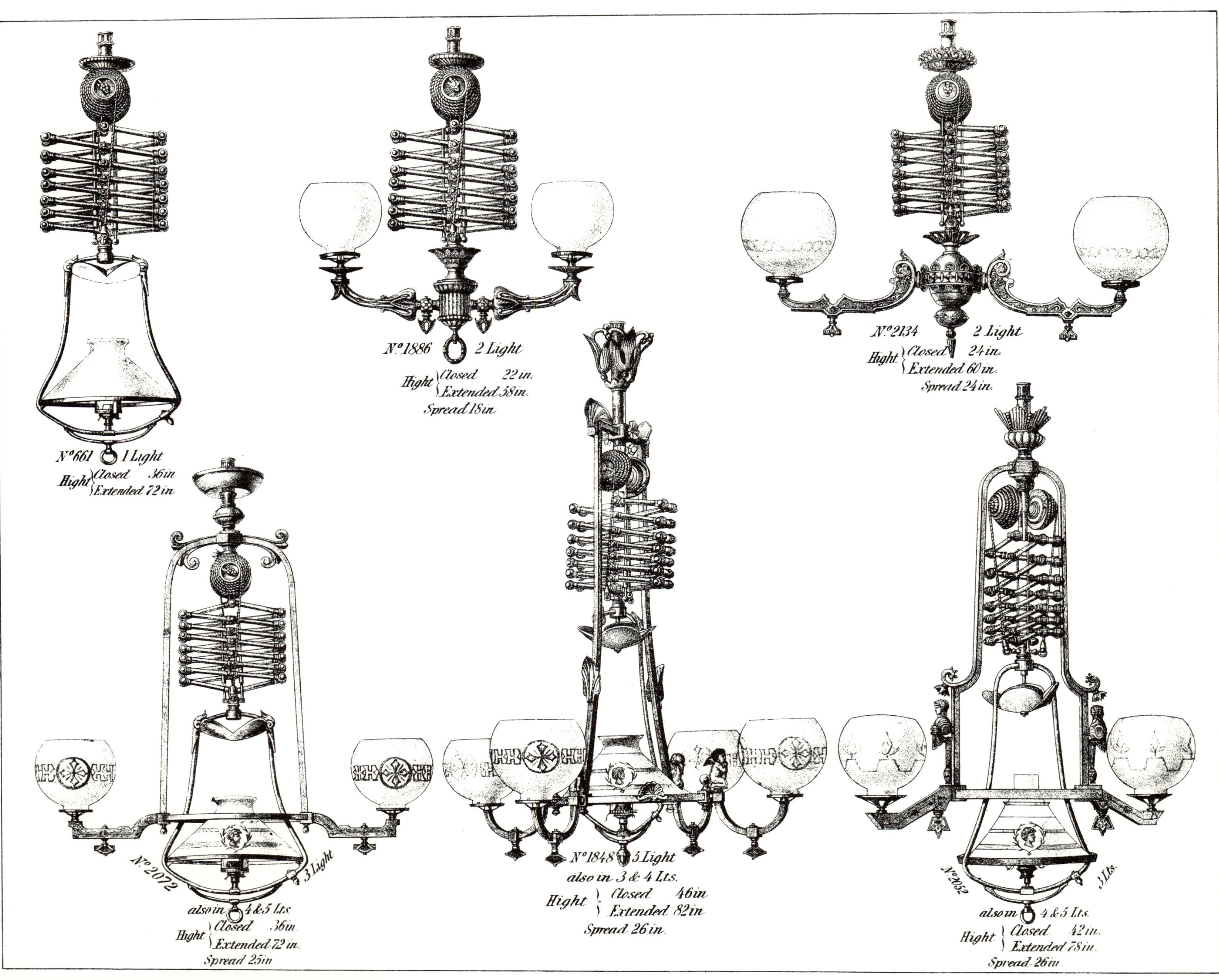

N° 661 1 Light
Hight Closed 36 in
Extended 72 in
N° 1886 2 Light
Hight Closed 22 in.
Extended 58 in.
Spread 18 in.
N° 2134 2 Light
Hight Closed 24 in.
Extended 60 in.
Spread 24 in.
N° 2072 3 Light
also in 4 & 5 Lts.
Hight Closed 36 in.
Extended 72 in.
Spread 25 in
N° 1848 5 Light
also in 3 & 4 Lts.
Hight Closed 46 in
Extended 82 in
Spread 26 in.
N° 2052 3 Lts.
also in 4 & 5 Lts.
Hight Closed 42 in.
Extended 78 in.
Spread 26 in

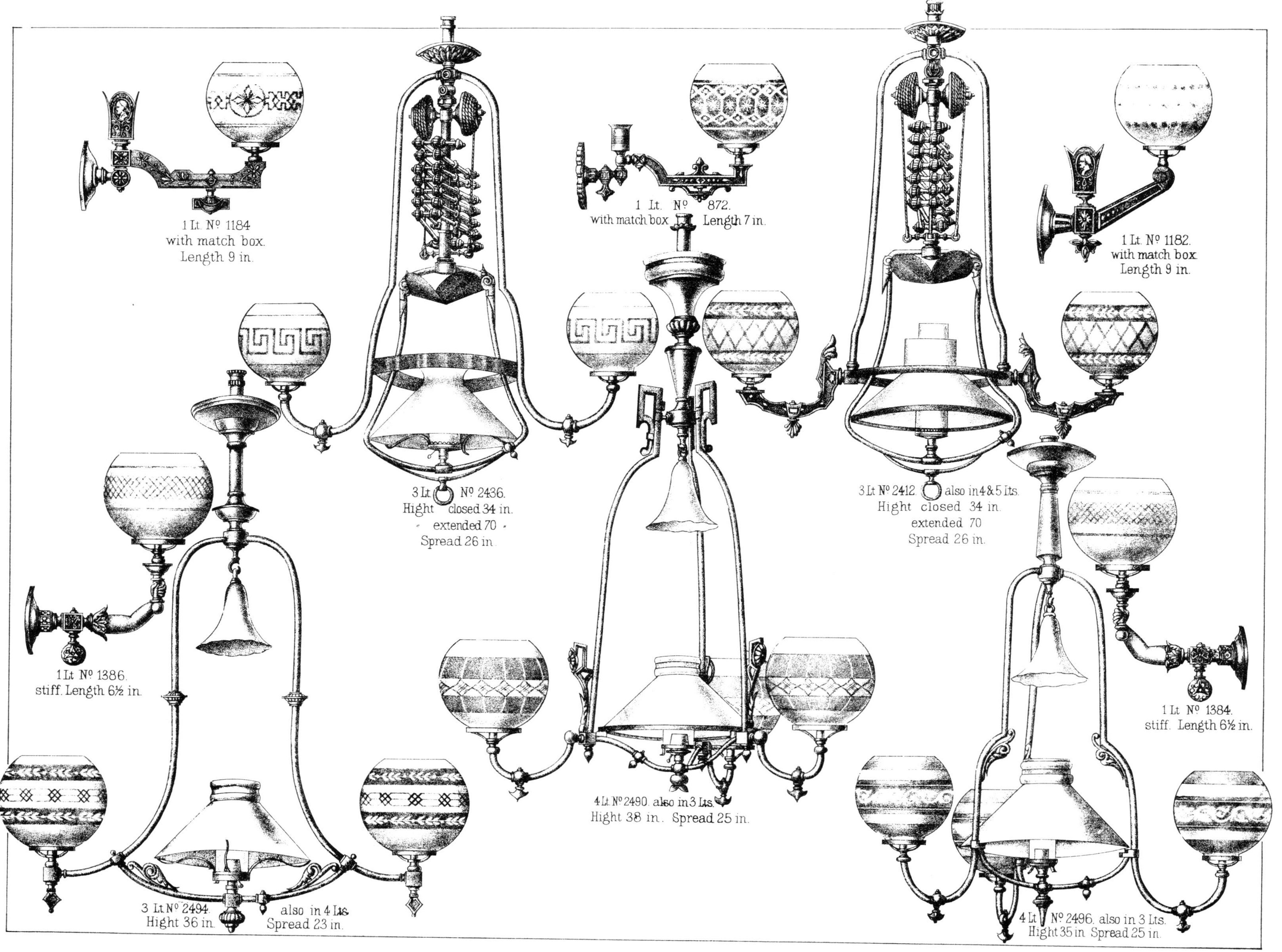
1 Lt. Nº 1184
with match box.
Length 9 in.
1 Lt. Nº 872.
with match box Length 7 in.
1 Lt. Nº 1182.
with match box
Length 9 in.
3 Lt. Nº 2436.
Hight closed 34 in.
" extended 70 "
Spread 26 in.
3 Lt. Nº 2412 also in 4 & 5 Lts.
Hight closed 34 in.
extended 70
Spread 26 in.
1 Lt. Nº 1386.
stiff. Length 6½ in.
1 Lt. Nº 1384.
stiff. Length 6½ in.
4 Lt. Nº 2490. also in 3 Lts.
Hight 38 in. Spread 25 in.
3 Lt. Nº 2494.
Hight 36 in.
also in 4 Lts.
Spread 23 in.
4 Lt. Nº 2496. also in 3 Lts.
Hight 35 in. Spread 25 in.

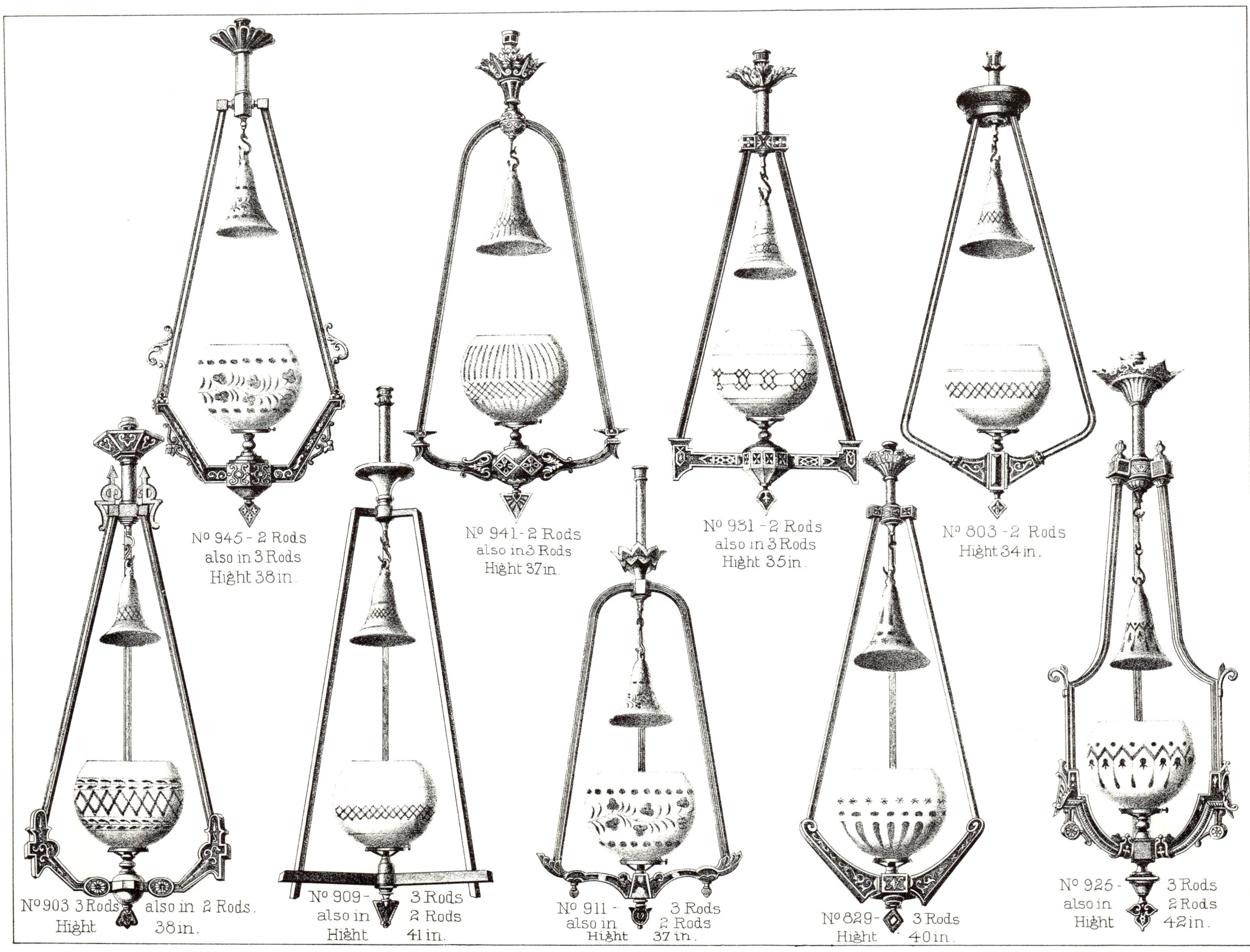
Nº 945 - 2 Rods
also in 3 Rods
Hight 38 in.
Nº 941 - 2 Rods
also in 3 Rods
Hight 37 in.
Nº 931 - 2 Rods
also in 3 Rods
Hight 35 in.
Nº 803 - 2 Rods
Hight 34 in.
Nº 903 3 Rods also in 2 Rods.
Hight 38 in.
Nº 909 - 3 Rods
also in 2 Rods
Hight 41 in.
Nº 911 - 3 Rods
also in 2 Rods
Hight 37 in.
Nº 829 - 3 Rods
Hight 40 in.
Nº 925 - 3 Rods
also in 2 Rods
Hight 42 in.

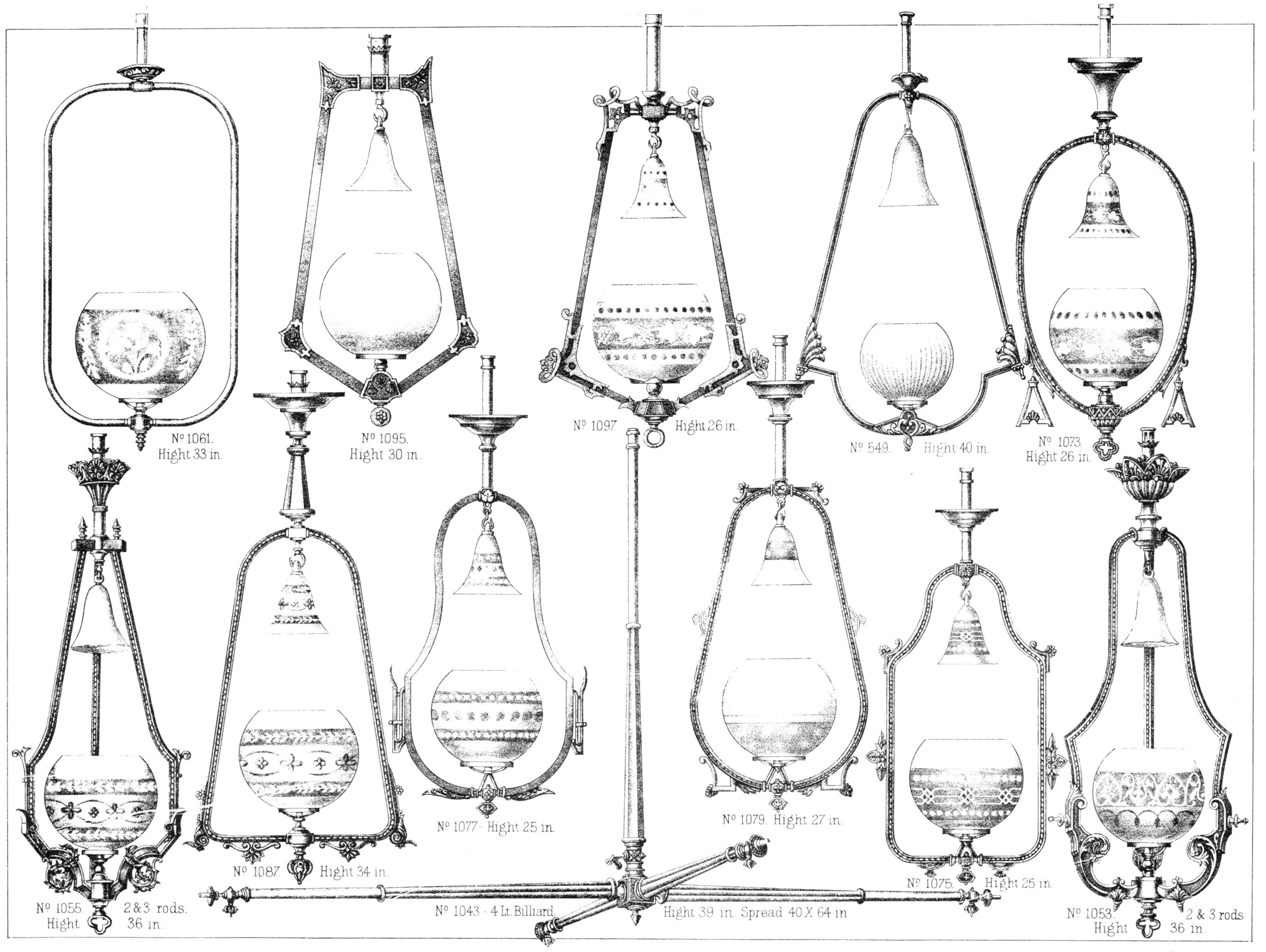
Nº 1061.
Hight 33 in.
Nº 1095.
Hight 30 in.
Nº 1097.
Hight 26 in.
Nº 549.
Hight 40 in.
Nº 1073.
Hight 26 in.
Nº 1055.
Hight
2 & 3 rods.
36 in.
Nº 1087.
Hight 34 in.
Nº 1077 - Hight 25 in.
Nº 1043 - 4 Lt. Billiard.
Hight 39 in. Spread 40 X 64 in.
Nº 1079. Hight 27 in.
Nº 1075.
Hight 25 in.
Nº 1053.
Hight
2 & 3 rods
36 in.

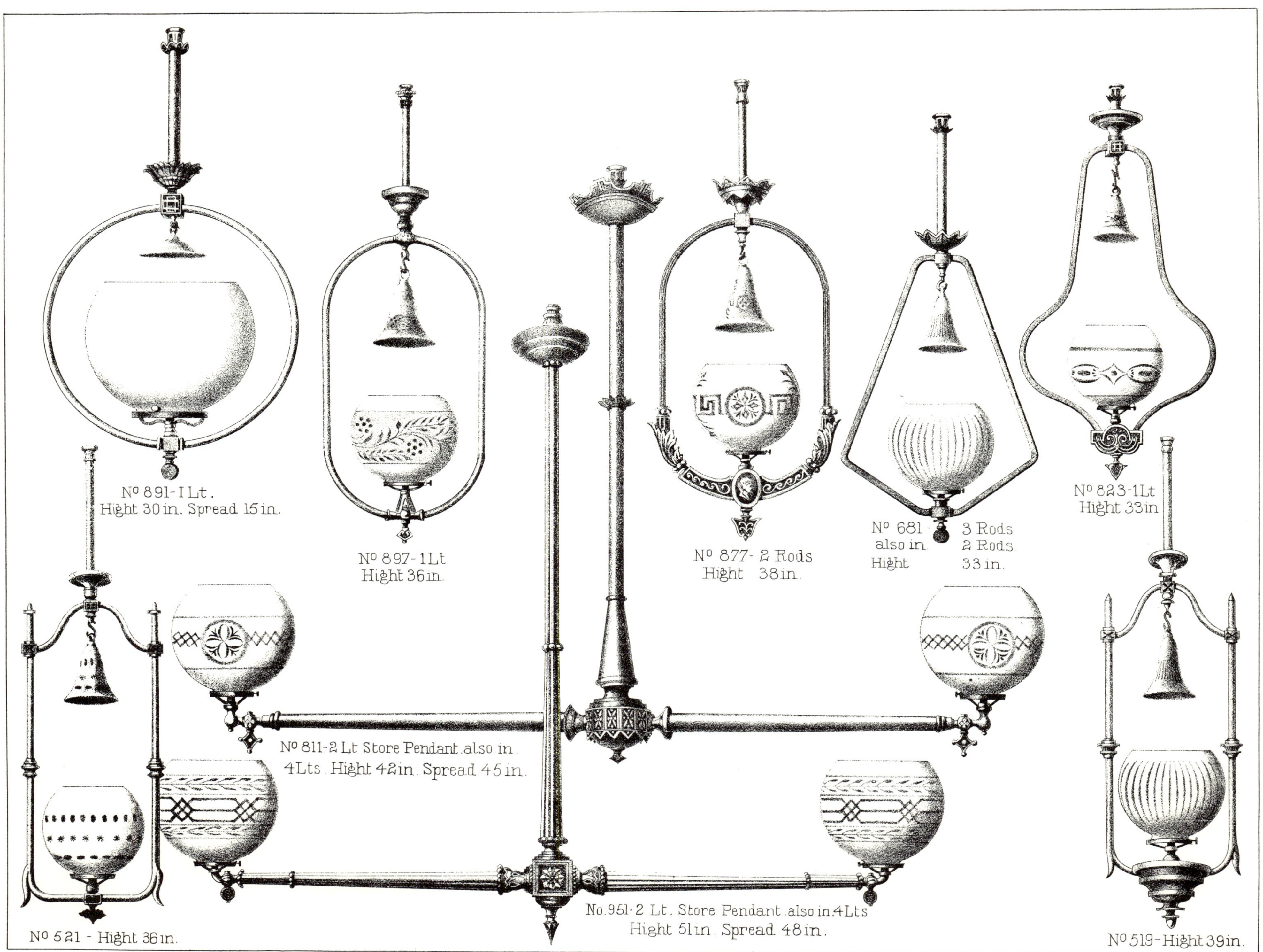

Nº 891-1 Lt.
Hight 30 in. Spread 15 in.
Nº 897-1 Lt
Hight 36 in.
Nº 877- 2 Rods
Hight 38 in.
Nº 681 - 3 Rods
also in 2 Rods
Hight 33 in.
Nº 823-1 Lt
Hight 33 in
Nº 811-2 Lt Store Pendant also in
4 Lts. Hight 42 in. Spread 45 in.
Nº 521 - Hight 36 in.
No. 951-2 Lt. Store Pendant also in 4 Lts
Hight 51 in. Spread 48 in.
Nº 519-Hight 39 in.

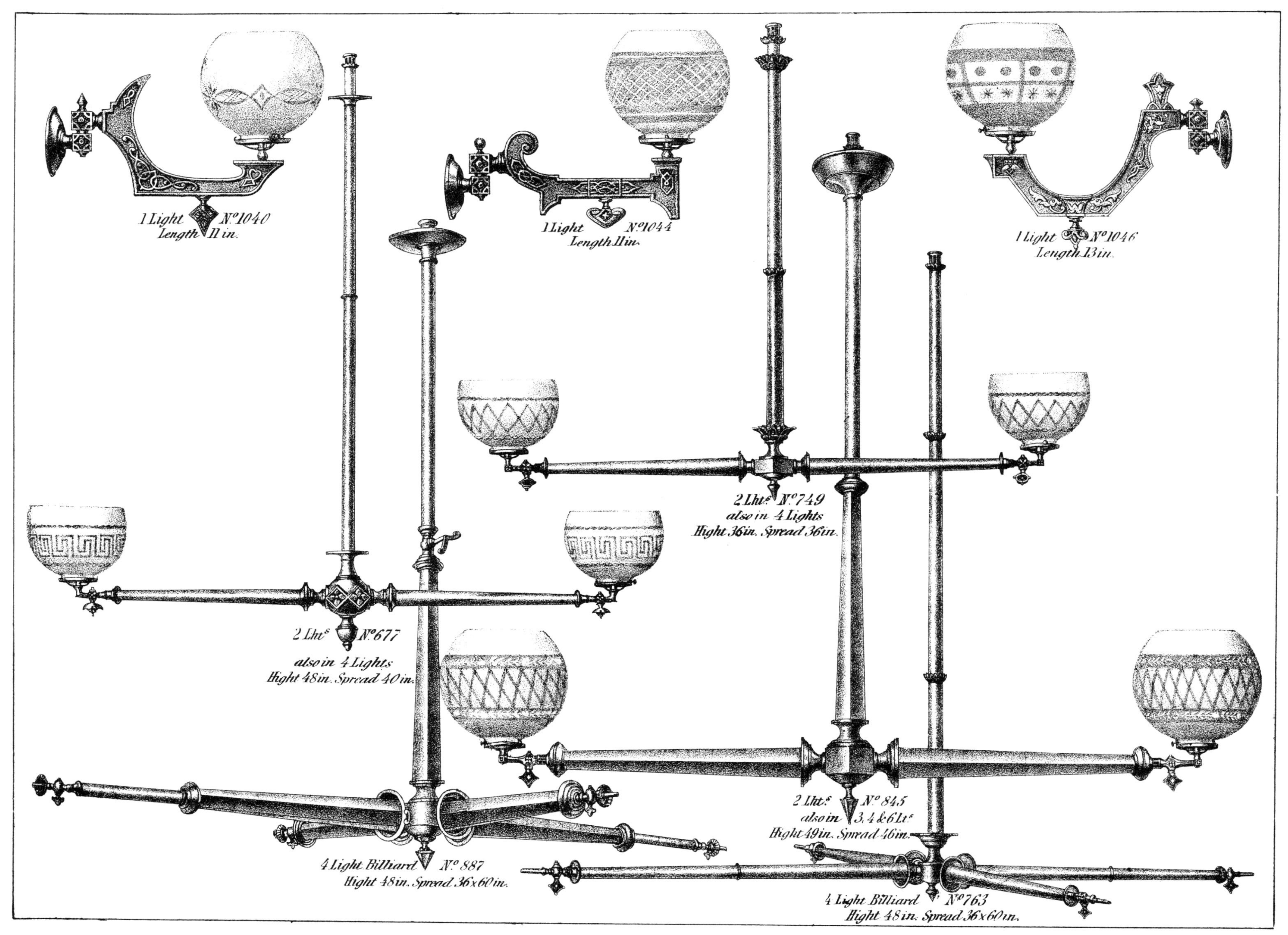
1 Light N.º 1040
Length 11 in.
1 Light N.º 1044
Length 11 in.
1 Light N.º 1046
Length 13 in.
2 Lht.s N.º 749
also in 4 Lights
Hight 36 in. Spread 36 in.
2 Lht.s N.º 677
also in 4 Lights
Hight 48 in. Spread 40 in.
2 Lht.s N.º 845
also in 3, 4 & 6 Lt.s
Hight 49 in. Spread 46 in.
4 Light Billiard N.º 887
Hight 48 in. Spread 36 x 60 in.
4 Light Billiard N.º 763
Hight 48 in. Spread 36 x 60 in.

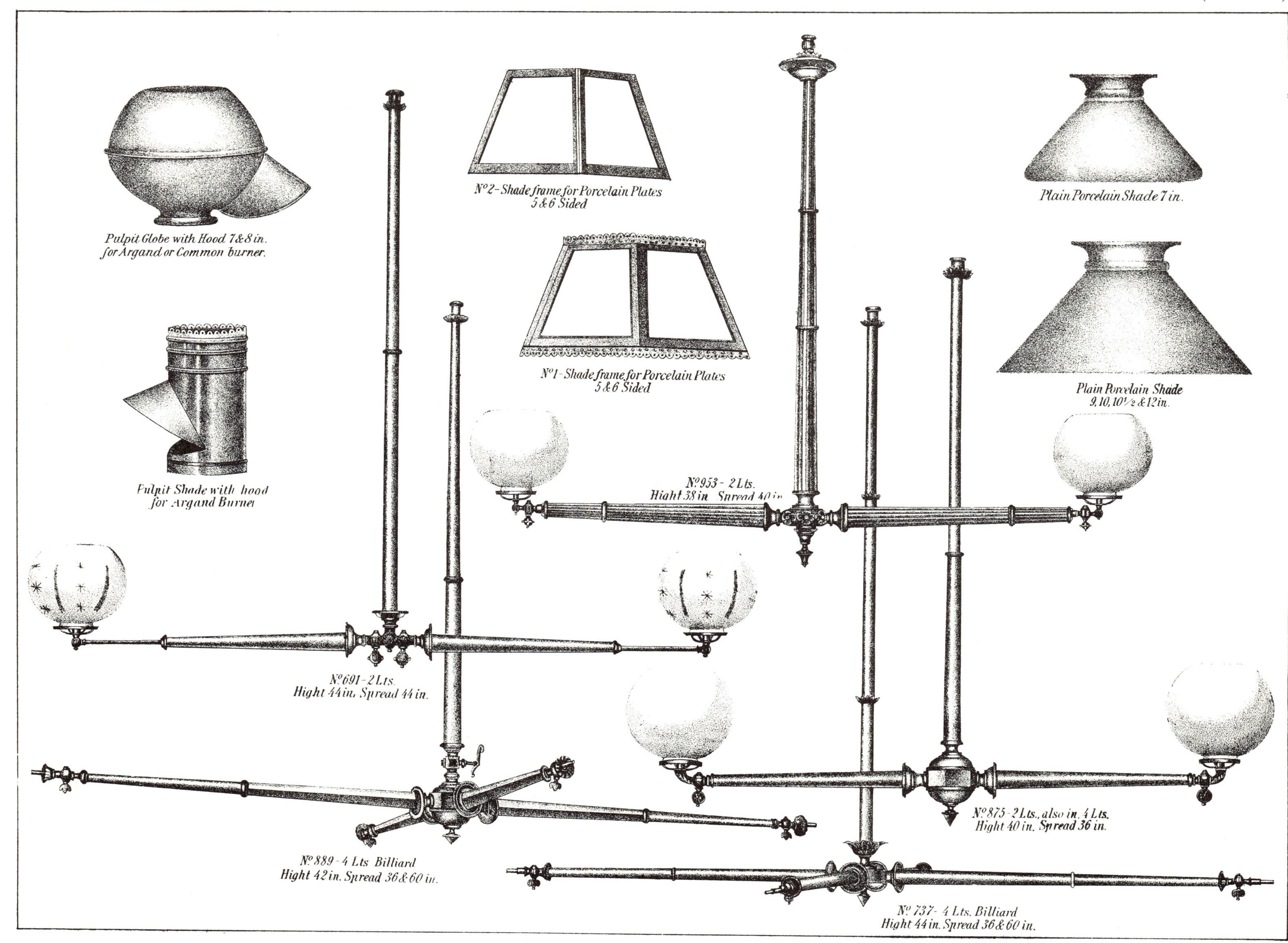

Pulpit Globe with Hood 7&8 in.
for Argand or Common burner.
Nº2 - Shade frame for Porcelain Plates
5 & 6 Sided
Plain Porcelain Shade 7 in.
Nº1 - Shade frame for Porcelain Plates
5 & 6 Sided
Plain Porcelain Shade
9, 10, 10½ & 12 in.
Pulpit Shade with hood
for Argand Burner
Nº953 - 2 Lts.
Hight 38 in. Spread 40 in.
Nº691 - 2 Lts.
Hight 44 in. Spread 44 in.
Nº875 - 2 Lts., also in 4 Lts.
Hight 40 in. Spread 36 in.
Nº889 - 4 Lts Billiard
Hight 42 in. Spread 36 & 60 in.
Nº737 - 4 Lts. Billiard
Hight 44 in. Spread 36 & 60 in.

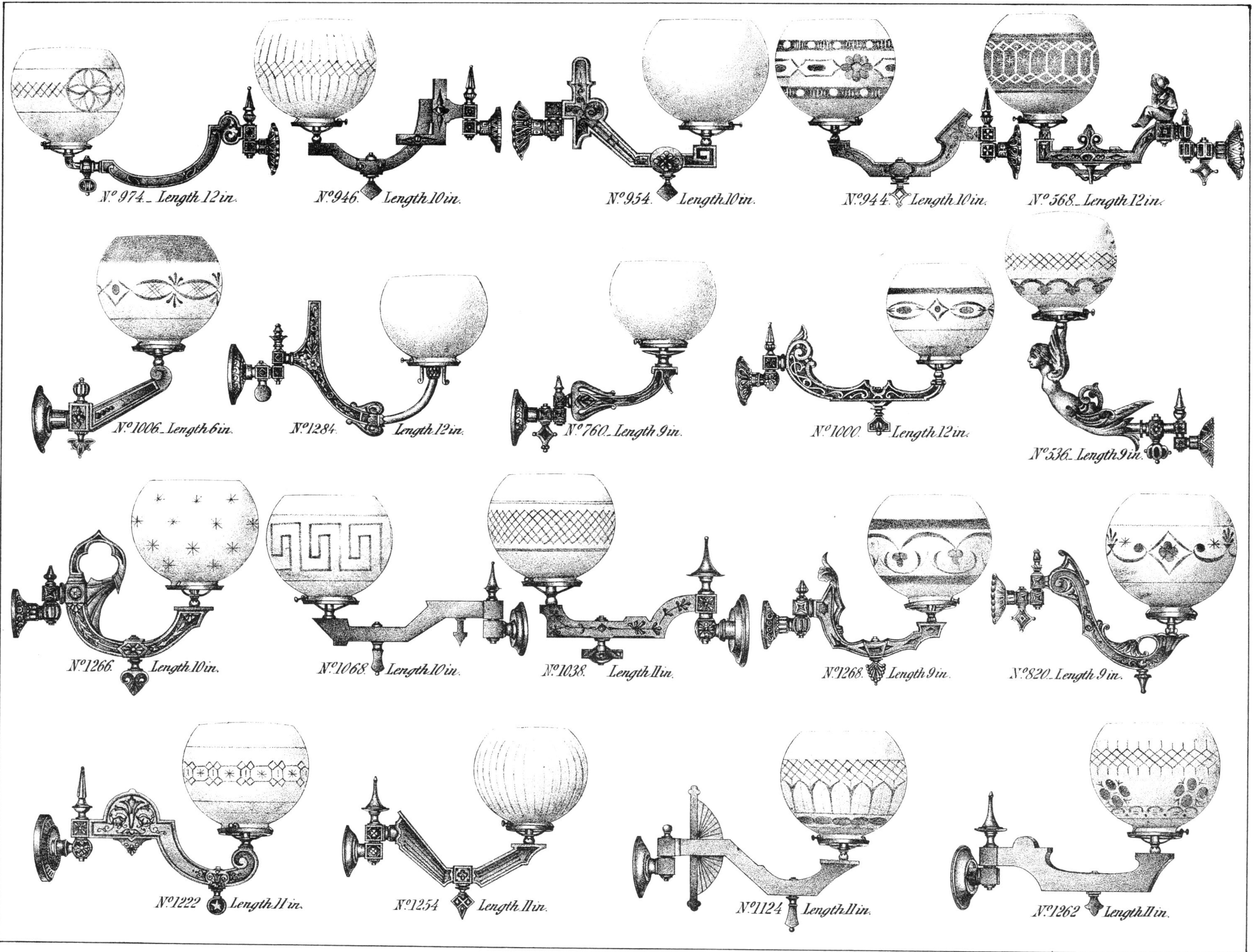
Nº 974._Length 12 in.
Nº 946. Length 10 in.
Nº 954. Length 10 in.
Nº 944. Length 10 in.
Nº 368._Length 12 in.
Nº 1006._Length 6 in.
Nº 1284. Length 12 in.
Nº 760._Length 9 in.
Nº 1000. Length 12 in.
Nº 536._Length 9 in.
Nº 1266. Length 10 in.
Nº 1068. Length 10 in.
Nº 1038. Length 11 in.
Nº 1268. Length 9 in.
Nº 820._Length 9 in.
Nº 1222 Length 11 in.
Nº 1254 Length 11 in.
Nº 1124 Length 11 in.
Nº 1262 Length 11 in.

No. 952
Spread 15 in.
No. 1282
Spread 14 in.
No. 1278
Spread 14 in.
No. 1070
Spread 15 in.
No. 914
Spread 17 in.
No. 1140
Spread 18 in.
No. 950
Spread 14 in.
No. 1276
Spread 15 in.
No. 1288.
Spread 11 in.
No. 956
Spread 15 in.
No. 1216
Spread 15 in.
No. 1264
Spread 17 in
Any of the above Brackets can be furnished in 3 Lights.

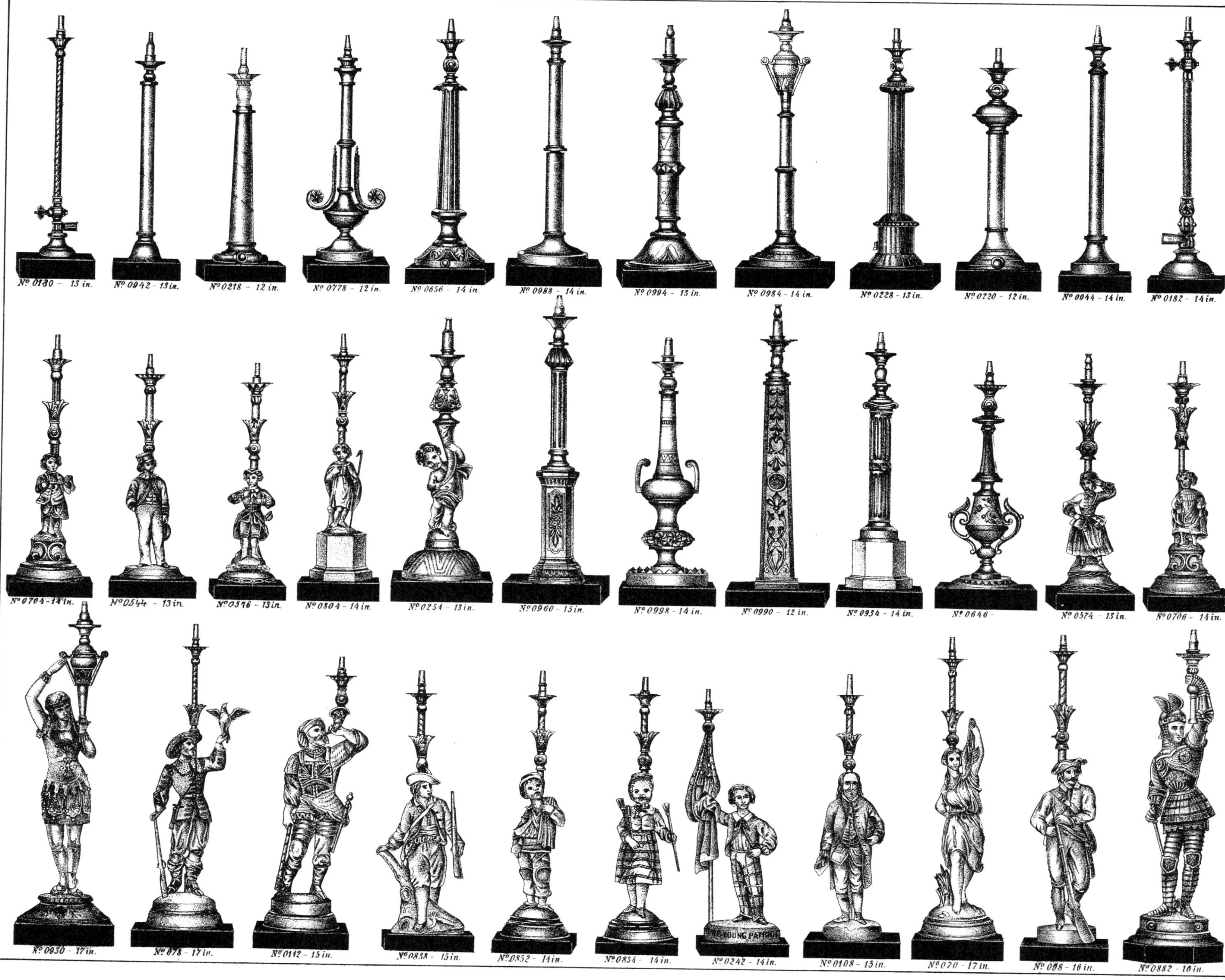
No 0180 – 13 in.
No 0942 – 13 in.
No 0218 – 12 in.
No 0778 – 12 in.
No 0656 – 14 in.
No 0988 – 14 in.
No 0994 – 13 in.
No 0984 – 14 in.
No 0228 – 13 in.
No 0220 – 12 in.
No 0944 – 14 in.
No 0182 – 14 in.
No 0704 – 14 in.
No 0544 – 13 in.
No 0576 – 13 in.
No 0804 – 14 in.
No 0254 – 13 in.
No 0960 – 15 in.
No 0998 – 14 in.
No 0990 – 12 in.
No 0934 – 14 in.
No 0646 –
No 0574 – 13 in.
No 0706 – 14 in.
No 0930 – 17 in.
No 078 – 17 in.
No 0112 – 15 in.
No 0838A – 13 in.
No 0852 – 14 in.
No 0854 – 14 in.
YOUNG PATRIOT
No 0242 – 14 in.
No 0108 – 13 in.
No 070 – 17 in.
No 0882 – 16 in.

Nº 0902 - 14in.
Nº 0950 - 14 in.
Nº 0534 - 10 in.
Nº 0970 - 7 in.
Nº 0968 - 18 in.
Nº 0560 - 12 in.
Nº 0536 - 12 in.
Nº 0962 - 21 in.
Nº 0792 - 8 in.
Nº 096 - 6 in.
Nº 0234 - 14 in.
Nº 0898 - 2 Lts. Hight 36 in. Spread 20 in.
Nº 0874 - 2 Lts. 2 joint Also in 1 joint Hight 24 in.
Nº 01036 - 2 Lts. 2 joint Also in 1 joint Hight 24 in.
Nº 0892 - 2 Lts. Hight 36 in. Spread 19 in.
Nº 0850 - 1 Lt. Hight 25 in.
Nº 0980 - 1 Lt. Hight 26 in.
Nº 0916 - 1 Lt. Hight 26 in.
Nº 0914 - 1 Lt. Hight 23 in.
Nº 0978 - 1 Lt. Hight 30 in.
Nº 0976 - 1 Lt. Hight 24 in.
Nº 0974 - 1 Lt. Hight 24 in.
Nº 0982 - 1 Lt. Hight 27 in.

No286
No288
No290
No292
No294
No296
No100
No108
No190
No202
No204
No212
No230
No236
No238
No240
No242
No244
No250
No252
No262
No266
No272
4, 5, 6 & 8 in.
Cut Bell
4, 5, 6 & 8 in.
Plain Bell

358.
352.
360.
344.
354.
356.
350.
336.
340.
332.
326.
346.
348.
324.
342.
328.
334.
318.
330
308.
314.
310.
312.
316.
338.
320.
322.
306.

No. 512.—Octagon.
(Small Corners.)
SIZES.

Hight.		Width.
34 in.	x	14 in.
40 "	x	16 "
44 "	x	20 "
61 "	x	23 "
68 "	x	28 "

No. 502.
Hexagon, Taper.
(Fancy Head.)
SIZES.

Hight.		Width.
36 in.	x	19 in.
42 "	x	23 "
50 "	x	27 "

No. 508.—Octagon.
(Square Top and Bottom.)
SIZES.

Hight.		Width.
40 in.	x	16 in.
48 "	x	22 "

No. 510.—Fancy Street Lantern.
SIZE.

Hight.		Width.
28 in.	x	15½ in.

No. 506.—Square.
(Reversed Top.)
SIZES.

Hight.		Width.
24 in.	x	13½ in.
27 "	x	16 "
31 "	x	19 "

No. 500.—Street Lantern.
SIZE.

Hight.		Width.
28 in.	x	15½ in.

No. 528.—The Prince Albert.
SIZE.

Hight.		Diam.
64 in.	x	24 in.

No. 504.—Fancy. *(Square.)*
SIZES.

Hight.		Width.
36 in.	x	14½ in.
42 "	x	16½ "
48 "	x	18½ "

No. 518.—Octagon. *(Small Corners, Taper.)*
SIZES.

Hight.		Width.
36 in.	x	18 in.
41 "	x	20 "
46 "	x	23 "
55 "	x	26 "

No. 514.—Square.
Reversed Top and Bottom.
SIZES.

Hight.		Width.
36 in.	x	21 in.
39 "	x	21½ "
45 "	x	24 "
55 "	x	27 "

No. 516.—Hexagon.
SIZE.

Hight.		Width.
45 in.	x	27 in.
54 "	x	30 "
60 "	x	33 "
66 "	x	36 "
78 "	x	42 "

No. 524.
Triangle Side Lantern.
SIZES.

Width.		Hight.
10 in.	x	16 in.
12 "	x	18 "
13 "	x	19 "

No. 522.
With Ornamental Bracket, No. 1.
SIZES.

Hight.		Diam.	Length of Bracket.
32	x	16 in.	30 in.
42	x	20 "	36 "
52	x	24 "	36 "

No. 526.—Side Lantern.
SIZE.

Width				Hight.
7	x	9 in.	x	16 in.
7½	x	12 "	x	18 "
10	x	12 "	x	18 "

Ground Globes.
10 to 14 inch.
With Ring Holder and Plain Canopy.

Ground Globes.
10 to 14 in.
With Ring Holder and Fancy Canopy.

No. 522.—12 Panel Globe Lantern.
SIZES.

Hight.		Diam.
32 in.	x	16 in.
42 "	x	20 "
52 "	x	24 "

No. 530.
SIZES.

Hight.		Diameter.
32	x	16 in.
42	x	20 "
52	x	24 "

No. 520.
8 Panel Globe Lantern.
SIZES.

Hight.		Diam.
32 in.	x	16 in.
42 "	x	20 "
52 "	x	24 "

No. 522.
With Ornamental Bracket,
No. 2.
SIZES.

Hight.		Diam.	Lenght of Bracket.
32	x	16 in.	30 in.
42	x	20 "	36 "
52	x	24 "	36 "

No. 1310 - 2 Lt.
also in 3 Lights Spread 19 in.

No. 1308 - 1 Lt.
Length 13 in.

No. 987 - 2 Lt.
also in 4 Lights
Hight 48 in. Spread 48 in.

No. 993 - 2 Lt.
also in 4 Lights
Hight 48 in. Spread 48 in.

No. 989 - 2 Lt.
also in 4 Lights
Hight 63 in. Spread 66 in.

No. 991 - 2 Lt.
Hight 72 in.
also in 4 & 6 Lights
Spread 60 in.

N° 01038 Hight 14 in
N° 01074 Hight 12 in.
N° 01062 Hight 12 in
N° 01060 Hight 13 in.
N° 01052 Hight 12 in
N° 01054 Hight 12 in
N° 01056 Hight 15 in.
N° 01050 Hight 14 in.
N° 2384-2 Lts
also in 3 & 4 Lts.
Hight 35 in. Spread 20 in.
N° 2362-2 Lts.
also in 3 & 4 Lts
Hight 36 in. Spread 21 in
N° 2350-2 Lts.
also in 3 & 4 Lts.
Hight 36 in. Spread 21 in
N° 5102 Match Box
Hight 9 in.
N° 2340-2 Lts
also in 3 & 4 Lts
Hight 38 in. Spread 22 in
N° 2360-2 Lts.
also in 3 & 4 Lts.
Hight 36 in. Spread 21 in
N° 01058-Hight 14 in.

Nº 1324-1 Lt.
Length 11 in.
Nº 1318-1 Lt.
Length 5 in.
Nº 1322-1 Lt.
Length 12 in.
Nº 1328-1 Lt.
Length 10 in.
Nº 1320-1 Lt.
Length 11 in.
Nº 1326-1 Lt. Bracket.
Length 7 in.
Nº 1072-1 Lt.
Hight 21 in.
Nº 2342-2 Lt.
also in 3,4&6 Lts.
Hight 42 in. Spread 25 in.
Nº 2356-2 Lt.
also in 3 & 4 Lts.
Hight 40 in. Spread 24 in.
Nº 2358-2 Lt. also in 3 & 4 Lts.
Hight 36 in. Spread 21 in.
Nº 2348-2 Lt. also in 3 & 4 Lts.
Hight 40 in. Spread 21 in.
Nº 2352-2 Lt. also in 3 & 4 Lts.
Hight 36 in. Spread 21 in.

Nº 1330-1 Lt - Gilt
Length 10½ in
Nº 1332-1 Lt - Gilt
Length 11½ in
Nº 1334-1 Lt - Gilt
Length 10½ in
Nº 2348-4 Lt
also in 2 & 3 Lts
Hight 40 in Spread 21 in
Nº 2352-4 Lt
also in 2 & 3 Lts
Hight 36 in Spread 21 in
Nº 2342-4 Lt also in 2,3 & 6 Lts
Hight 42 in Spread 25 in
Nº 2340-4 Lt also in 2 & 3 Lts
Hight 38 in Spread 22 in
Nº 2356-4 Lt also in 2 & 3 Lts
Hight 40 in Spread 24 in

N° 2368 - 3 Lts.
also in 2 & 4 Lts.
Hight 23 in Spread 11 in
N° 01066 Hight 17 in
N° 01064 Hight 17 in
N° 1336 - 3 Lts.
also in 1, 2 & 4 Lts.
Total Hight 33 in
to burner
N° 1003 - 2 Rods
Hight 34 in.
N° 2362 - 4 Lts.
also in 2 & 3 Lts.
Hight 36 in. Spread 21 in
N° 2350 - 4 Lts.
also in 2 & 3 Lts.
Hight 36 in. Spread 21 in.
N° 2358 - 4 Lts.
also in 2 & 3 Lts. Hight 36 in Spread 21 in
N° 2370 - 4 Lts. also in 2, 3 & 6 Lts
Hight 46 in. Spread 30 in
N° 2360 - 4 Lts. also in 2 & 3 Lts Hight 36 in Spread 21 in

Nº 0744
Hight 16 in
Nº 01006.
Hight 15 in.
Nº 0732.
Hight 16 in.
Nº 2380-4 Lts. also in 3 & 6 Lts. Hight 48 in Spread 32 in.
Nº 2382 6 Lts. also in 3 & 4 Lts. Hight 48 in Spread 34 in.
Nº 875-2 Lts. also in 4 Lts.
Hight 40 in Spread 36 in.
Nº 995-2 Lts. also in 4 Lts.
Hight 40 in Spread 36 in
Nº 2378 4 Lts. also in 2, 3 & 6 Lts.
Hight 42 in Spread 25 in
Nº 1001-2 Lts also in 4 Lts Hight 42 in Spread 42 in.

THE DOME GAS STOVES

WILL NOT PRODUCE

Any Deleterious or Disagreeable Odor.

THE DOME GAS COOKING STOVES are warranted to do all kinds of cooking, possible with coal or wood, at less expense, and without smoke or any disagreeable odor.

DIRECTIONS FOR USE.

Hold a lighted match just above the burner when the gas is turned on, that none may escape.

Always light the gas before ANY UTENSIL is placed on top of the stove.

In using the oven, light the gas, place the oven on the stove, and let it get thoroughly hot before placing anything in it to bake. It will take from three to five minutes. In using the broiler, place your meat in the broiler before putting it on the stove.

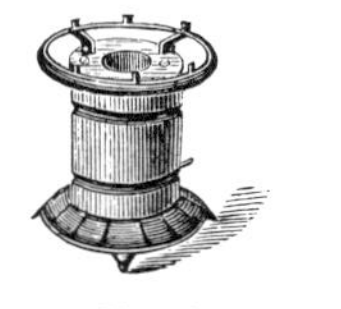
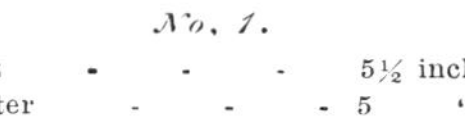

No. 1.

Height - - - 5½ inches.
Diameter - - - 5 "

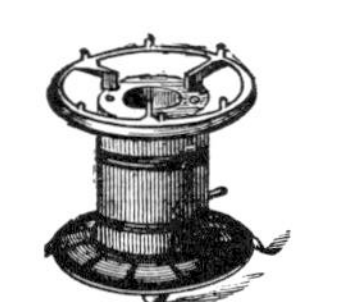

No. 2.

Height, - - - 6 inches.
Diameter, - - - 6 "

Nos. 1 and 2 are intended more especially for TABLE USE, nursery purposes, or sick rooms, or for heating small sad-irons, and are attached to the gas burner by flexible tubing.

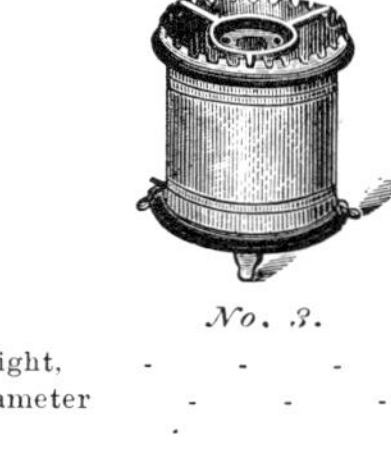

No. 3.

Height. - - - 7 inches.
Diameter - - - 7½ "

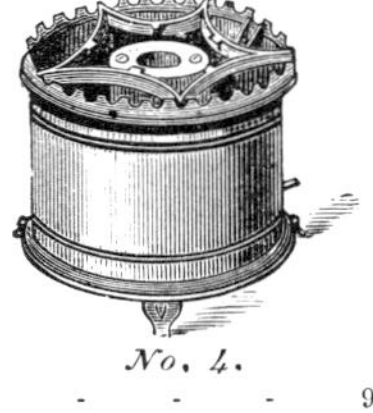

No. 4.

Height, - - - 9 inches.
Diameter, - - - 8 "

A sufficient flow of gas may usually be obtained through an open burner for either Nos. 3 or 4.

The No. 1 oven and both broilers will fit and work perfectly over either, if a sufficient flow of gas is allowed.

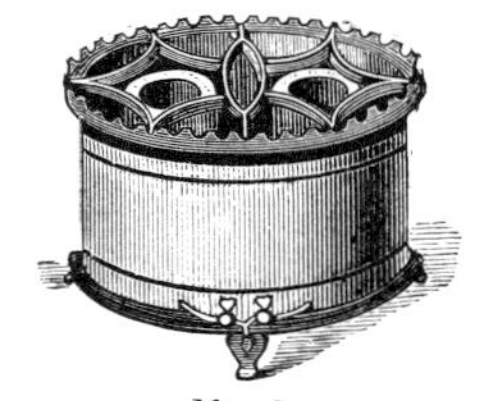

No. 5.

Oval—Height, 9 inches. Length, 11 inches.

This size is especially adapted to laundry use. As it will not deposit soot, sad-irons can be brought in immediate contact with the flame, and not be deprived of any heat by intervening plates, as in other stoves.

The No. 5 has two burners, supplied by a double gas-cock, by which one or both may be used at pleasure. For boiling considerable quantities of water, broiling large steaks, etc., heating flat-irons, a tailor's goose, or a griddle, it will be found very useful.

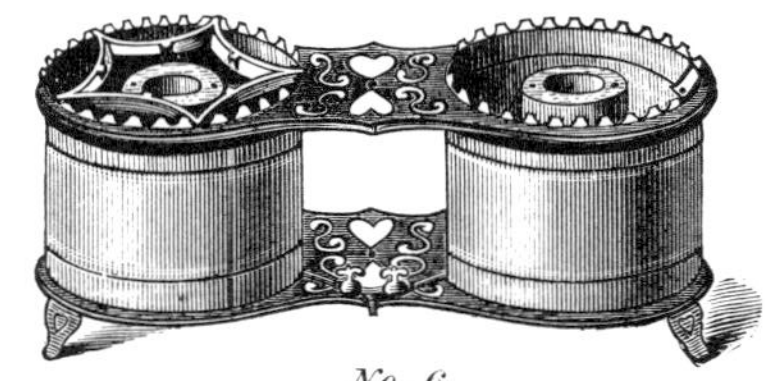
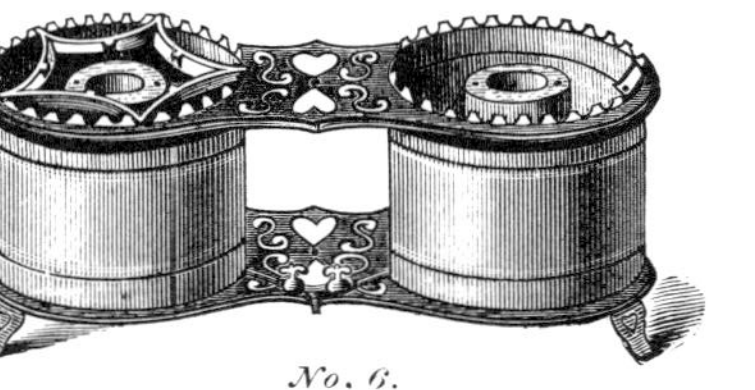

No. 6.

Height, 9 inches. Length, 21 inches.

No. 6 is a double stove, supplied like No. 5, through a double gas-cock, enabling the operator to light one or both at pleasure. With this stove, cooking may be done for a family of five or six persons, and at much less expense than with coal and wood, as the fire may be extinguished, and the expense cease, the instant the cooking is completed. It will do any work reasonably expected of an ordinary stove.

Patented July 14, 1868.

DOME GAS OVEN AND BROILER.

TWO SIZES OF EACH.

No. 1 Oven, 12 in. wide, 14 in long.
" 2 " 14 " " 20 " "
" 0 Broiler, 10 " x 10 "
" 1 " 10 " x 12 "

These utensils are guaranteed to work to perfection over the same Gas Stoves. The No. 1 Oven fits the Nos. 3, 4, 6, or 7. The No. 2 only the 5 or 7. The broilers fit all sizes. The steaks stand upright, and cook both sides at once, and no gravy is lost.

This is the Finest Heating Stove ever made.

FOUR SIZES.

No. 12—2 B's. | No. 14—4 B's.
" 13—3 " | " 15—5 "

These Heaters give perfect satisfaction, and are very convenient for all rooms where there are no chimney flues. They make no smoke, smell, dust, or ashes.

No. 7.

Length and depth, 21 inches.

This Stove has three holes and four burners, and in short, is Nos. 5 and 6 combined, and must always be connected with iron pipe to insure the best results.

It will do all the cooking and laundry work of a family of eight or twelve persons. This is the largest size Dome Stove made, except to order.

Gas Ætna,

FOR NURSERY PURPOSES,
HEATING WATER, MAKING COFFEE, TEA, GRUEL, ETC.

Will accommodate any vessel, from a small tin-cup to a good sized coffee-pot.

Works perfectly over an ordinary fish-tail burner. Will not *smoke* or *light down*, and is the best article for the purpose ever offered to the public.

Excelsior Gas Ætna.

This is a burner we have just introduced. It is the lowest Ætna, with a perfect combustion, that was ever made, and is destined to take the place of all others.

THE ÆTNA

GAS STOVE.

THIS STOVE, next to the Dome, is the very best one for all purposes ever offered to the public. The burners will not smoke or light down, and are positively odorless, and are so arranged as to require no cleaning—never clogging up or burning out.

THE LINE OF

ÆTNA STOVES

Is extended sufficiently to adapt them to all the various uses required of a Gas Stove, either for family use or the trade.

We would call especial ttention to our new style of pattern of the Ætna Gas Stove; while the burner is the same as the old style, the pattern and construction are such as to make it, next to the Dome, the handsomest and best one for all purposes ever offered to the public.

Nos. 10, 20 and 30 equal in size the old Nos. 50, 60, and 80, while No. 32 equals the old No. 280, and No. 33 that of No. 384.

We take particular pains that the quality and pattern of our goods shall be such as to please the public and ensure demand.

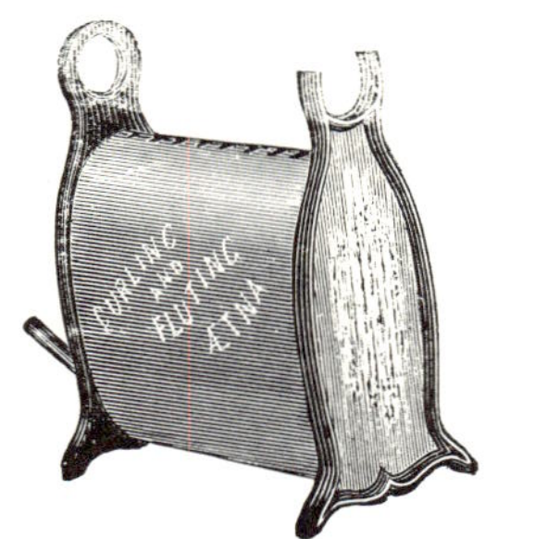

4 inches.
6 "

This cut represents an arrangement for heating FLUTING OR CURLING IRONS with gas. It is made with cast-iron ends and Russia-iron body, and will accommodate two irons at once.

The principle is the same as in the gas stoves, and the combustion is so perfect that no soot is deposited by the flame.

They are neat, cleanly, and always ready. The consumption of gas under full pressure is only *four* feet per hour.

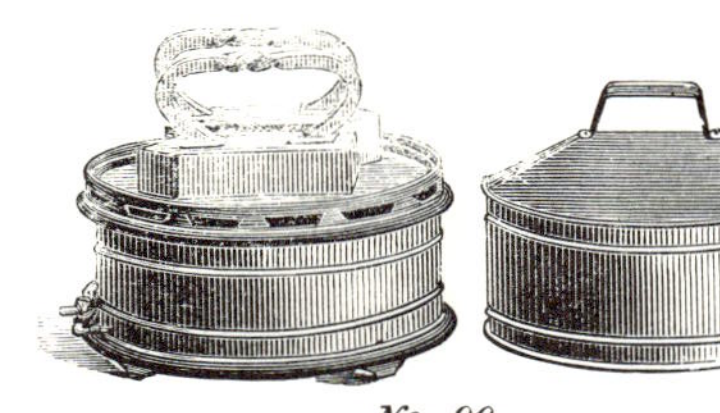

No. 90.

This cut represents a stove for tailors' use. It has four (4) burners, and is capable of heating two (2) geese at once. It can be used with 2 or 4 burners.

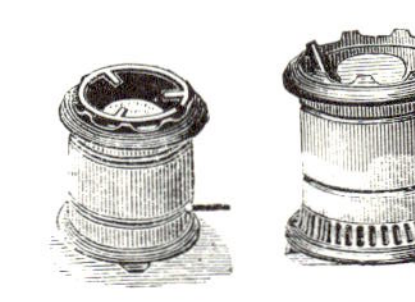

No. 40. *No. 10.* *No. 20.*

These sizes are intended more especially for nursery purposes, heating small sad irons, and for light cooking, and are attached to the gas burner by flexible tubing.

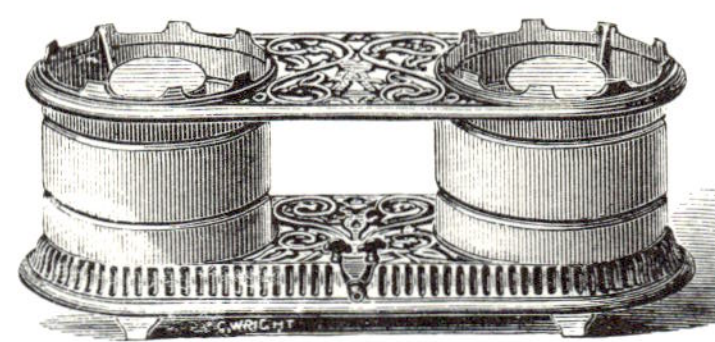

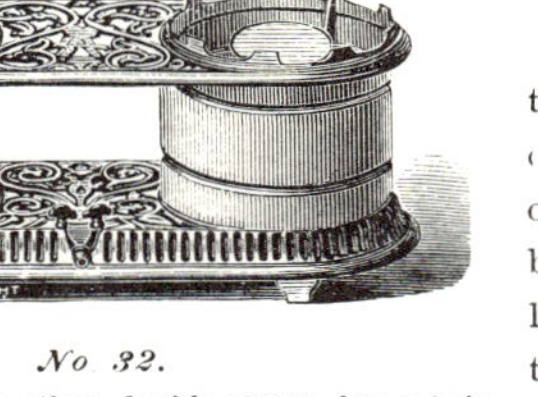

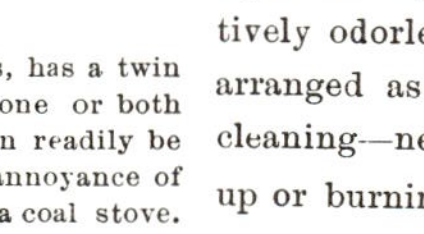

No. 32.

This size, like the other double stoves, has a twin cock, enabling the operator to light one or both sides. The cooking for a small family can readily be done on this size, thus avoiding the annoyance of heat and dust that is unavoidable with a coal stove.

No. 96.

This is a laundry stove, with six (6) burners for heating sad-irons. It is a splendid stove for summer use. Six irons can be heated at once.

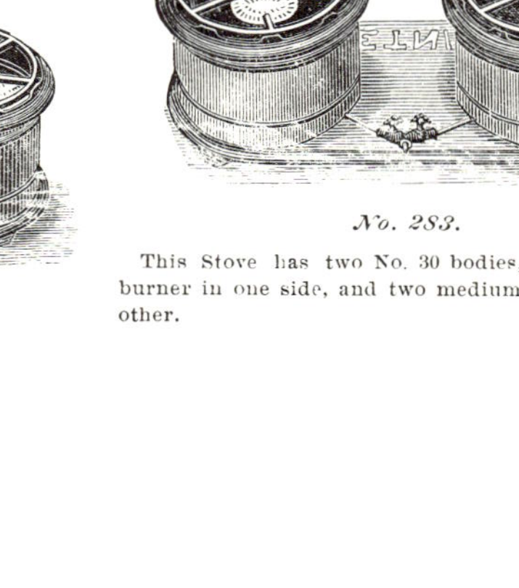

No. 30. *No. 82.*

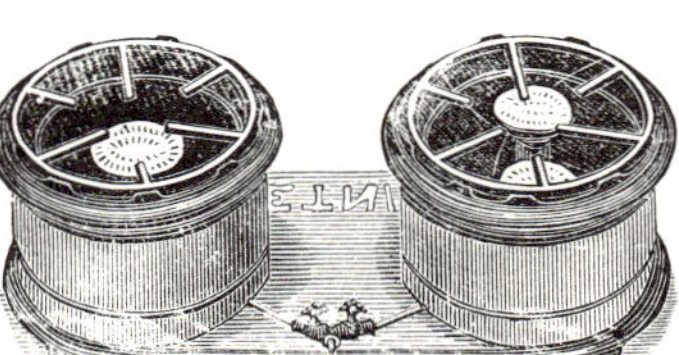

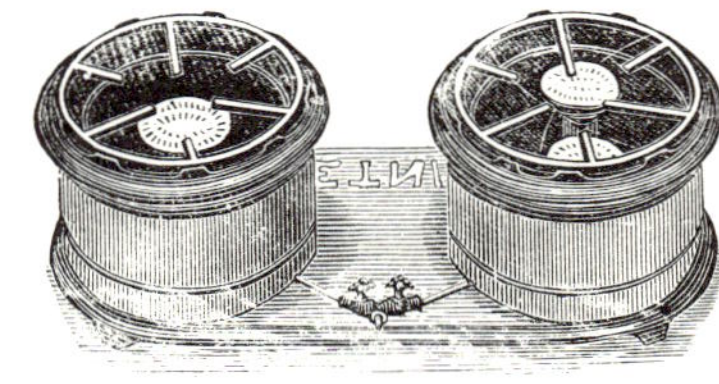

No. 283.

This Stove has two No. 30 bodies, with one large burner in one side, and two medium burners in the other.

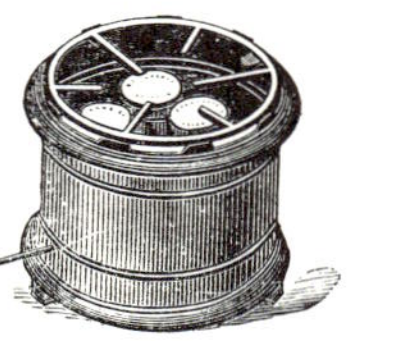

No. 83.

The numbers above represent the largest single-bodied Ætna Stoves, and are 8 inches diameter. The No. 30 has a large burner and is used for ordinary family use. The No. 82 has two medium burners in same sized body as No. 30. The No. 83 has three medium burners in same body. These are for use where an extra heat is desired in small compass, such as quickly cooking oysters, etc., etc.

The No. 82 will thoroughly heat a No. 2 Dome Oven, and is a very desirable size for the No. 1 Broiler.

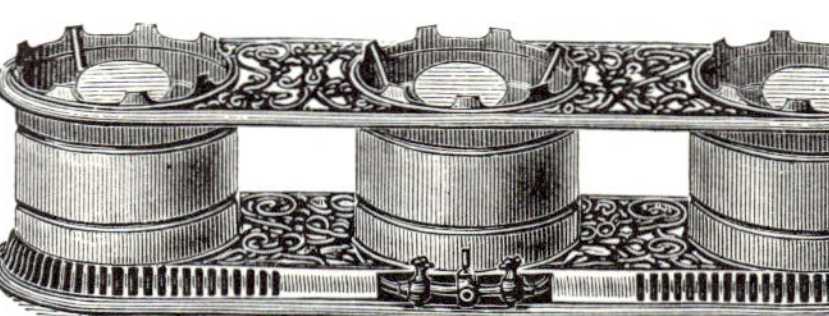

No. 33.

This size, the largest of the Ætna Stoves, is well adapted to cooking in large families, restaurants, and oyster saloons, and will do as much work as an ordinary summer range. The No. 1 Oven and either Broiler can be used on Nos. 20, 30, 32, or 33.

Ærated Gas Ætna,

FOR NURSERY PURPOSES.

HEATING WATER, MAKING COFFEE, TEAS, GRUEL, ETC.

Will accommodate any vessel, from a small tin cup to a good sized coffee pot.

Works perfectly over an ordinary fish-tail burner. Will not *smoke* or *light down*, and is the best article, for the purpose, ever offered to the public.

No. 1. *No. 2.*

TEA KETTLES.

No. 1.—5 in. Diam.	4 in. Deep.
" 2.—6 " "	5 " "

2 Lt Nº 1432.
Spread 15 in.
3 Lt. Nº 1438.
Spread 18½ in.
Nº 1051. 2 & 3 rods.
Hight 42 in.
Nº 1063. 2&3 rods.
Hight 46 in.
6 Lt Nº 2556.
also in 3, 4 & 12 Lts.
Ht. 3,4&6 Lts. 54 in Spd 32 in
Ht. 12 Lt. 60 in Spd 34 in
6 Lt. Nº 2512. also in 3 & 4 Lts.
Hight 45 in. Spread 28 in.
6 Lt. Nº 2536
also in 2,3 & 4 Lts Hight 45 in. Spread 32 in.

Portable Nº 01096.
Hight 17 in
Portable Nº 01098.
Hight 17 in.
Portable Nº 01116.
Hight 18 in.
3 Lt. Nº 01186 with cigar lighter attachmt
01186½ without
also in 4 & 5 Lts.
Hight 37 in. Spread 19½ in.
6 Lt. Nº 2526. also in 2,3 & 4 Lts.
Hight 42 in. Spread 26 in.
3 Lt. Nº 01182 with cigar lighter attchmt
01182½ without
also in 4 & 5 Lts.
Hight 41 in. Spread 22 in.
Portable Nº 01118.
Hight 18 in.
6 Lt. Nº 2530. also in 2,3 &
Hight 44 in. Spread 28
6 Lt. Nº 2522. also in 2,3 & 4 Lts.
Hight 50 in. Spread 32 in.

No 1132. 3 Lts.
also in 2 Lts. Spread 22 in.
2 Lt. No 1470.
Spread 14 in.
2 Lt. No 1444.
Spread 18 in.
Portable No 01204.
Hight 13½ in.
Cigar lighter to burn Alcohol.
No 01152. Hight 8 in
Cigar lighters to burn Alcohol.
No 01154. Ht 8 in. No 01156 Ht 8 in.
Sliding Portable No 01108.
Hight 20 in.
4 Lt. No 2534. also in 2,3 & 6 Lts.
Hight 44 in. Spread 27 in.
4 Lt. No 2524.
also in 2,3 & 6 Lts.
Hight 37 in. Spd 29 in.
6 Lt. No 2540. also in 2,3 & 4 Lts.
Hight 54 in. Spread 27 in.

3 Lt. also in 2 Lts. Nº 1402. Spread 24 in.
2 Lt. Nº 1430. also in 3 Lts. Spread 16 in.
3 Lt. Nº 1300. also in 2 Lts. Spread 21 in.
Portable 01032. Hight 12 in.
4 Lt. Nº 2514 also in 2, 3 & 6 Lts. Hight 36 in. {Spread in 2, 3 & 4 Lts 21 in 6 - 24.
4 Lt. Nº 2532. also in 2, 3 & 6 Lts Hight 44 in. Spread 26 in.
Portable 01160. Ht. 12 in.
Portable 01206. Hight 14 in.
2 Lt Store Pendt Nº 1019. also in 4 & 6 Lts Hight 52 in. Spread 52 in.
Portable 01202. Hight 14½ in.

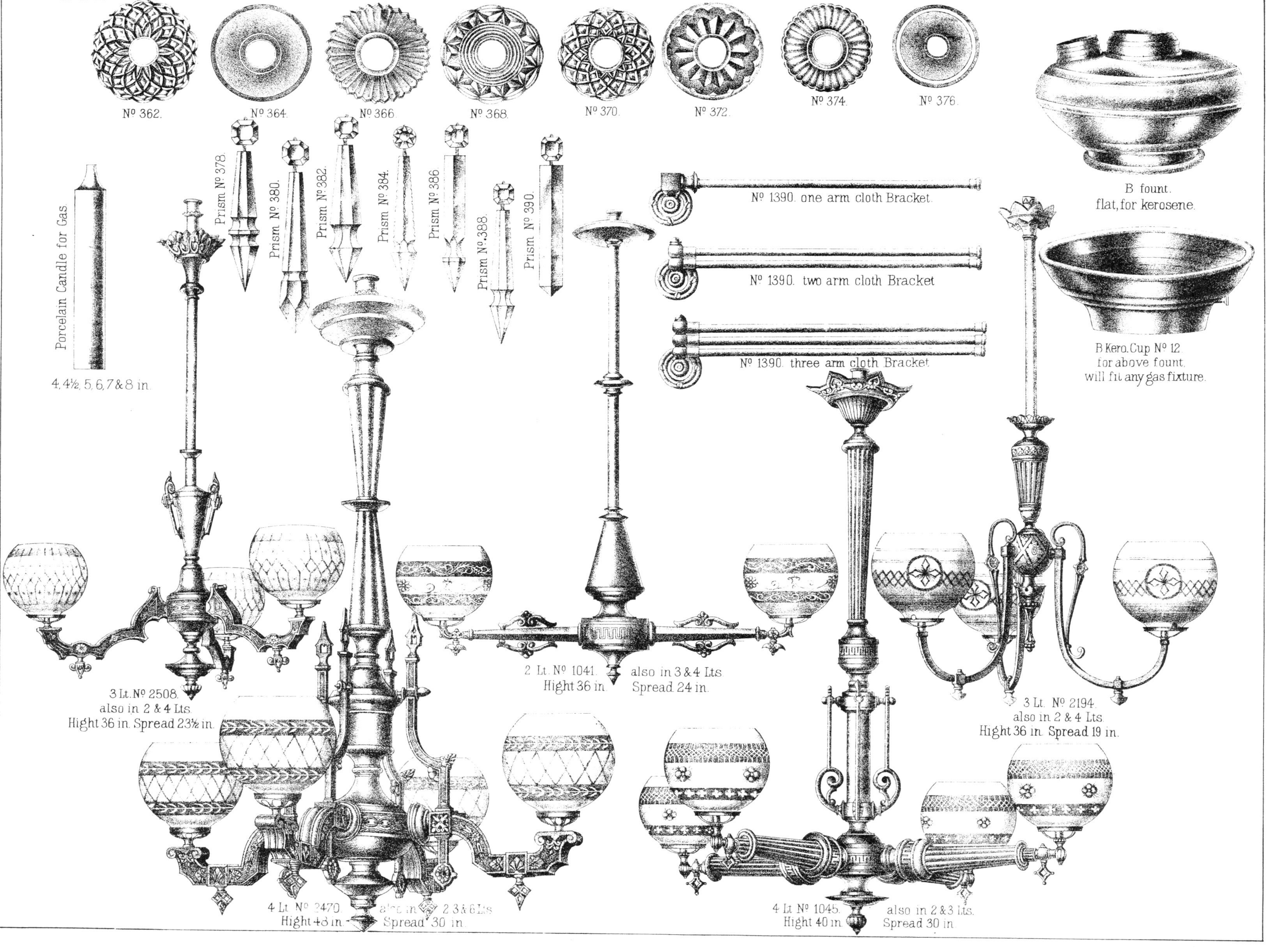
Nº 362.
Nº 364.
Nº 366
Nº 368.
Nº 370.
Nº 372.
Nº 374.
Nº 376.
B fount.
flat, for kerosene.
B Kero. Cup Nº 12
for above fount.
will fit any gas fixture.
Porcelain Candle for Gas.
4, 4½, 5, 6, 7 & 8 in.
Prism Nº 378.
Prism Nº 380.
Prism Nº 382.
Prism Nº 384.
Prism Nº 386
Prism Nº 388.
Prism Nº 390.
Nº 1390. one arm cloth Bracket.
Nº 1390. two arm cloth Bracket
Nº 1390. three arm cloth Bracket
3 Lt. Nº 2508.
also in 2 & 4 Lts.
Hight 36 in. Spread 23½ in.
2 Lt. Nº 1041.
Hight 36 in.
also in 3 & 4 Lts.
Spread 24 in.
3 Lt. Nº 2194.
also in 2 & 4 Lts.
Hight 36 in. Spread 19 in.
4 Lt. Nº 2470.
Hight 48 in.
Spread 30 in.
4 Lt. Nº 1045.
Hight 40 in.
also in 2 & 3 Lts.
Spread 30 in.

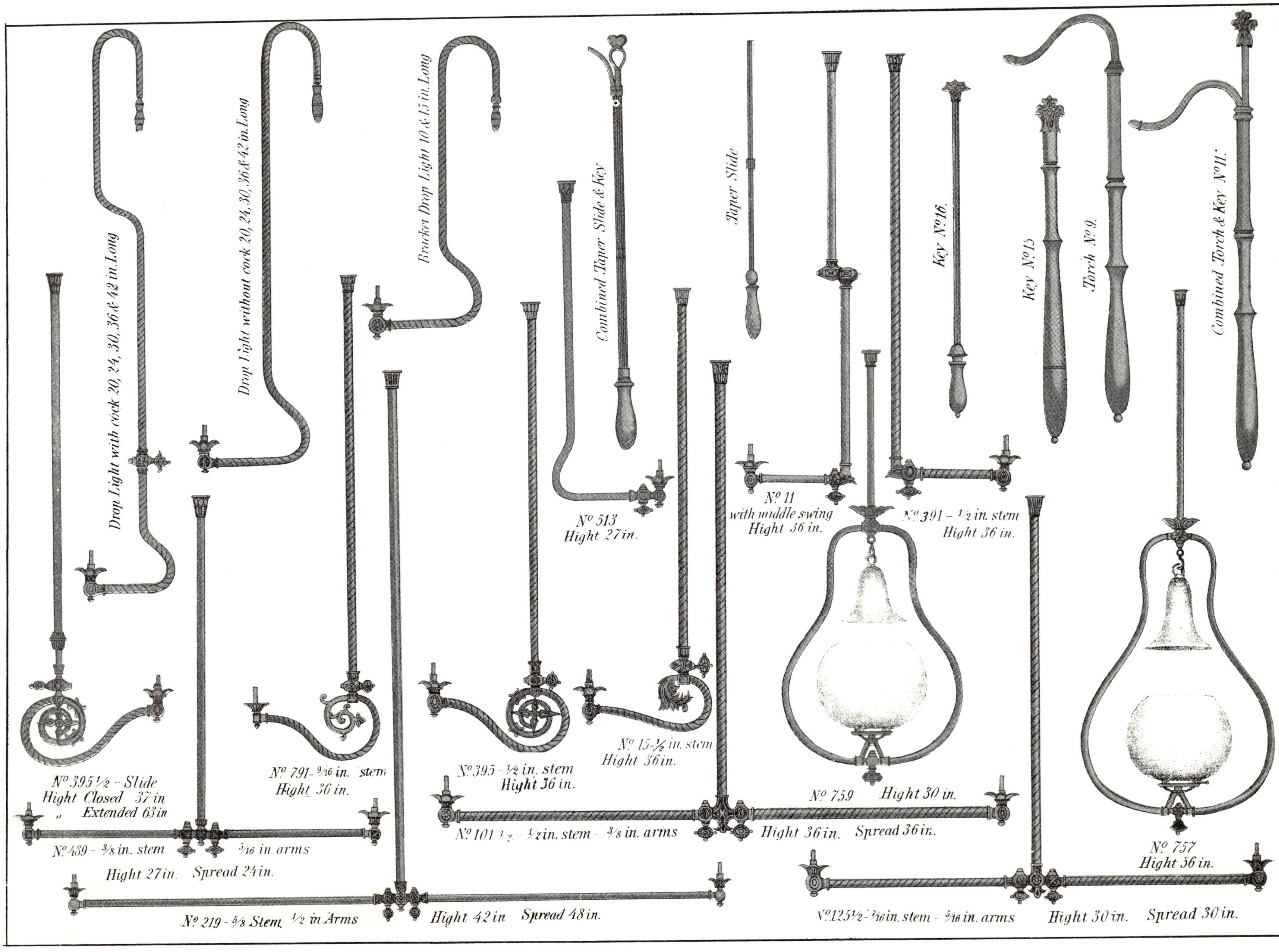

Drop Light with cock 20, 24, 30, 36 & 42 in. Long
Drop Light without cock 20, 24, 30, 36 & 42 in. Long
Bracket Drop Light 10 & 15 in. Long
Combined Taper Slide & Key
Taper Slide
Key Nº16.
Key Nº15
Torch Nº 9.
Combined Torch & Key Nº11.
Nº 513
Hight 27 in.
Nº 11
with middle swing
Hight 36 in.
Nº 391 - 1/2 in. stem
Hight 36 in.
Nº 395 1/2 - Slide
Hight Closed 37 in
" Extended 63 in
Nº 791 - 9/16 in. stem
Hight 36 in.
Nº 395 - 1/2 in. stem
Hight 36 in.
Nº 15 - 1/2 in. stem
Hight 36 in.
Nº 759 Hight 30 in.
Nº 101 1/2 - 1/2 in. stem - 3/8 in. arms
Hight 36 in. Spread 36 in.
Nº 489 - 3/8 in. stem 5/16 in. arms
Hight 27 in. Spread 24 in.
Nº 219 - 5/8 Stem 1/2 in Arms
Hight 42 in Spread 48 in.
Nº 757
Hight 36 in.
Nº 125 1/2 - 3/16 in. stem - 5/16 in. arms
Hight 30 in. Spread 30 in.

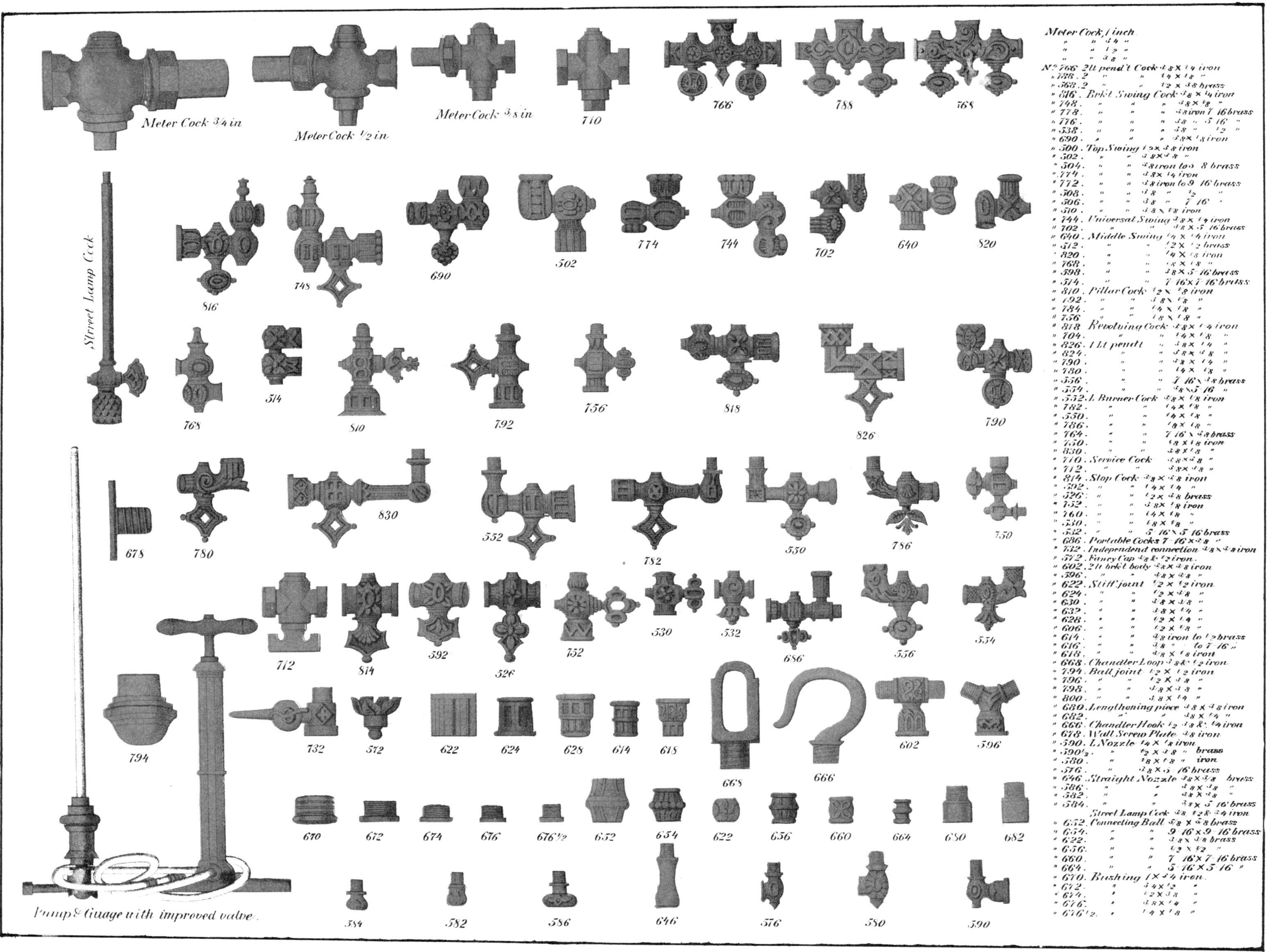
Meter Cock 3/4 in
Meter Cock 1/2 in
Meter Cock 3/8 in
710
766
788
768
Street Lamp Cock
816
748
690
502
774
744
702
640
820
768
514
810
792
756
818
826
790
678
780
830
552
782
550
786
750
712
814
592
526
752
530
532
686
556
534
794
732
572
622
624
628
614
618
668
666
602
596
670
672
674
676
676 1/2
652
654
622
656
660
664
650
682
Pump & Guage with improved valve.
584
582
586
646
576
580
590
Meter Cock, 1 inch
" " 3/4 "
" " 1/2 "
" " 3/8 "
No. 766 2 lt pend't Cock 3/8 × 1/4 iron
" 788 . 2 " " 1/4 × 1/8 "
" 568 . 2 " " 1/2 × 3/8 brass
" 816 . Brk't Swing Cock 3/8 × 1/4 iron
" 748 . " " " 3/8 × 1/8 "
" 778 . " " " 3/8 iron 7/16 brass
" 776 . " " " 3/8 " 5/16 "
" 538 . " " " 3/8 " 1/2 "
" 690 . " " " 3/8 × 1/8 iron
" 500 . Top Swing 1/2 × 3/8 iron
" 502 . " " 3/8 × 3/8 "
" 504 . " " 3/8 iron to 5/8 brass
" 774 . " " 3/8 × 1/4 iron
" 772 . " " 3/8 iron to 9/16 brass
" 508 . " " 3/8 " 1/2 "
" 506 . " " 3/8 " 7/16 "
" 510 . " " 3/8 × 1/8 iron
" 744 . Universal Swing 3/8 × 1/4 iron
" 702 . " " 3/8 × 5/16 brass
" 640 . Middle Swing 1/4 × 1/4 iron
" 512 . " " 1/2 × 1/2 brass
" 820 . " " 1/4 × 1/8 iron
" 768 . " " 1/8 × 1/8 "
" 598 . " " 3/8 × 5/16 brass
" 514 . " " 7/16 × 7/16 brass
" 810 . Pillar Cock 1/2 × 1/8 iron
" 792 . " " 3/8 × 1/8 "
" 784 . " " 1/4 × 1/8 "
" 756 " " 1/8 × 1/8 "
" 818 Revolving Cock 3/8 × 1/4 iron
" 704 . " " 1/4 × 1/8 "
" 826 . 1 Lt pendt " 3/8 × 1/4 "
" 824 . " " 3/8 × 3/8 "
" 790 . " " 3/8 × 1/4 "
" 780 . " " 1/4 × 1/8 "
" 556 . " " 7/16 × 3/8 brass
" 554 . " " 3/8 × 5/16 "
" 552 . L Burner Cock 3/8 × 1/8 iron
" 782 . " " 1/4 × 1/8 "
" 550 . " " 1/4 × 1/8 "
" 786 . " " 1/8 × 1/8 "
" 764 . " " 7/16 × 3/8 brass
" 750 . " " 1/8 × 1/8 iron
" 830 . " " 3/8 × 1/8 "
" 710 . Service Cock 3/8 × 3/8 "
" 712 . " " 3/8 × 3/8 "
" 814 . Stop Cock 3/8 × 3/8 iron
" 592 . " " 1/4 × 1/4 "
" 526 . " " 1/2 × 3/8 brass
" 752 . " " 3/8 × 1/8 iron
" 760 . " " 1/4 × 1/8 "
" 530 . " " 1/8 × 1/8 "
" 532 . " " 5/16 × 5/16 brass
" 686 . Portable Cocks 7/16 × 3/8 "
" 732 . Independend connection 3/8 × 3/8 iron
" 572 . Fancy Cap 3/8 & 1/2 iron.
" 602 . 2 lt brk't body 3/8 × 3/8 iron
" 596 . " " 3/8 × 3/8 "
" 622 . Stiff joint 1/2 × 1/2 iron
" 624 . " " 1/2 × 3/8 "
" 630 . " " 3/8 × 3/8 "
" 632 . " " 3/8 × 1/4 "
" 628 . " " 1/2 × 1/4 "
" 606 . " " 1/2 × 1/8 "
" 614 . " " 3/8 iron to 1/2 brass
" 616 . " " 3/8 " to 7/16 "
" 618 . " " 3/8 × 1/8 iron
" 668 . Chandler Loop 3/8 & 1/2 iron
" 794 . Ball joint 1/2 × 1/2 iron
" 796 . " " 1/2 × 3/8 "
" 798 . " " 3/8 × 3/8 "
" 800 . " " 3/8 × 1/4 "
" 680 . Lengthening piece 3/8 × 3/8 iron
" 682 . " " 3/8 × 1/4 "
" 666 . Chandler Hook 1/2 3/8 & 1/4 iron
" 678 . Wall Screw Plate 3/8 iron
" 590 . L Nozzle 1/4 × 1/8 iron
" 590 1/2 . " 1/2 × 3/8 " brass
" 580 . " 1/8 × 1/8 " iron
" 576 . " 3/8 × 5/16 brass
" 646 . Straight Nozzle 3/8 × 3/8 brass
" 586 . " " 3/8 × 3/8 "
" 582 . " " 3/8 × 3/8 "
" 584 . " " 3/8 × 5/16 brass
Street Lamp Cock 3/8, 1/2 & 3/4 iron
" 652 . Connecting Ball 5/8 × 5/8 brass
" 654 . " " 9/16 × 9/16 brass
" 622 . " " 3/8 × 3/8 brass
" 656 . " " 1/2 × 1/2 "
" 660 . " " 7/16 × 7/16 brass
" 664 . " " 5/16 × 5/16 "
" 670 . Bushing 1 × 3/4 iron.
" 672 . " 3/4 × 1/2 "
" 674 . " 1/2 × 3/8 "
" 676 . " 3/8 × 1/4 "
" 676 1/2 . " 1/4 × 1/8 "

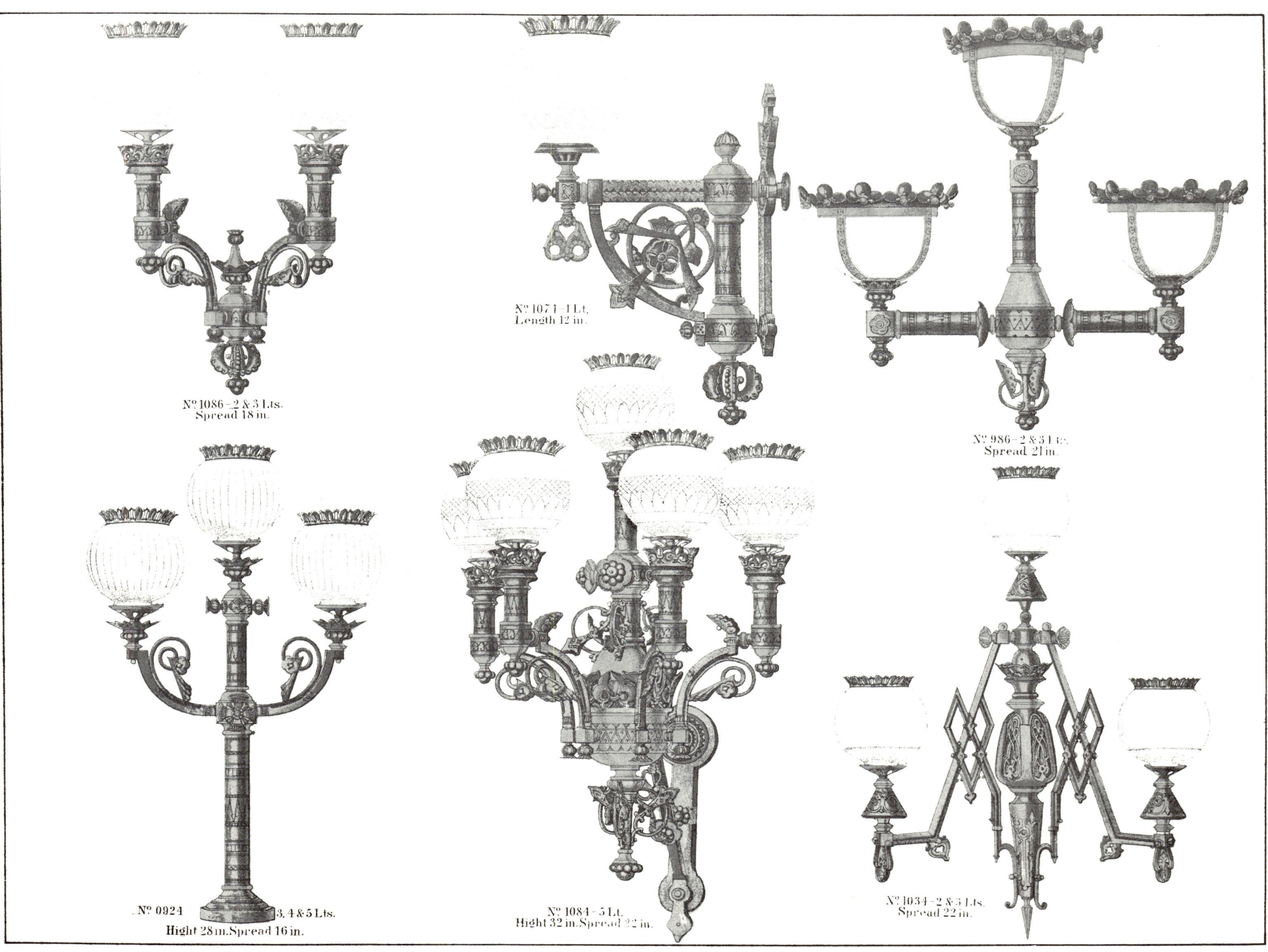

Nº 1086–2 & 3 Lts.
Spread 18 in.
Nº 1071–1 Lt.
Length 12 in.
Nº 986–2 & 3 Lts.
Spread 21 in.
Nº 0924 3, 4 & 5 Lts.
Hight 28 in. Spread 16 in.
Nº 1084–5 Lt.
Hight 32 in. Spread 22 in.
Nº 1034–2 & 3 Lts.
Spread 22 in.

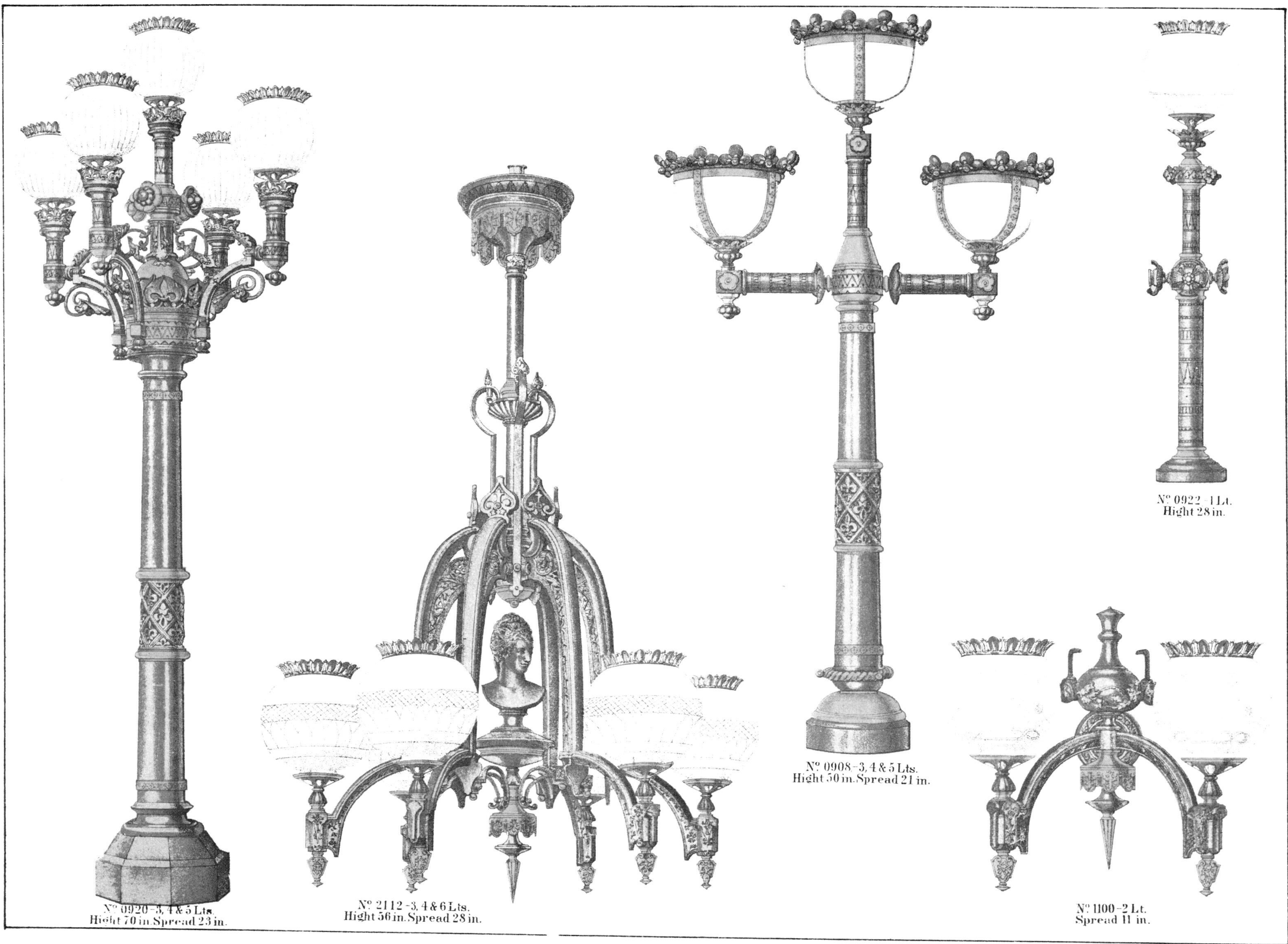
Nº 0920–3, 4 & 5 Lts.
Hight 70 in. Spread 23 in.
Nº 2112–3, 4 & 6 Lts.
Hight 56 in. Spread 28 in.
Nº 0908–3, 4 & 5 Lts.
Hight 50 in. Spread 21 in.
Nº 0922–1 Lt.
Hight 28 in.
Nº 1100–2 Lt.
Spread 11 in.

No. 2098 - 8, 12, 16, 18, 24 & 32 Lts. Chandelier. Hight 7 3/12 ft. Spread 5 8/12 ft.

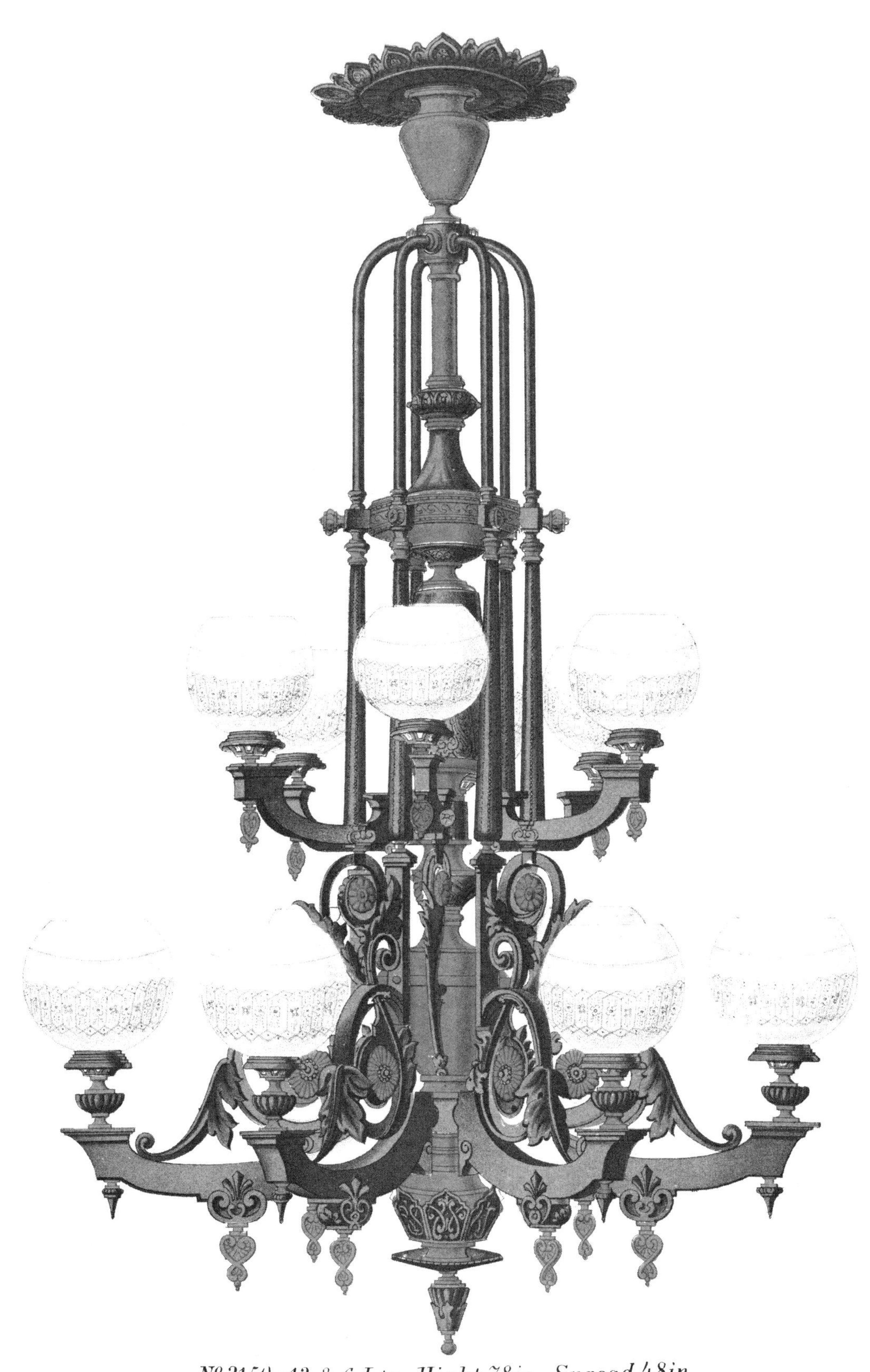

Nº 2150 - 12 & 6 Lts. Hight 78in. Spread 48in.

Nº 2200 - 36, 50 & 64 Lts. Hight 9 ft. Spread 6 2/12 ft.

No. 100
No. 104
No. 106
No. 108
No. 110
No. 112
No. 114
No. 116
No. 118
No. 120
No. 124
No. 126
No. 130
No. 134
No. 140
No. 142
No. 148
No. 150
No. 152
No. 154
Metal Coronet for Globes. Gilt or Bronze
No. 156
No. 158
No. 160

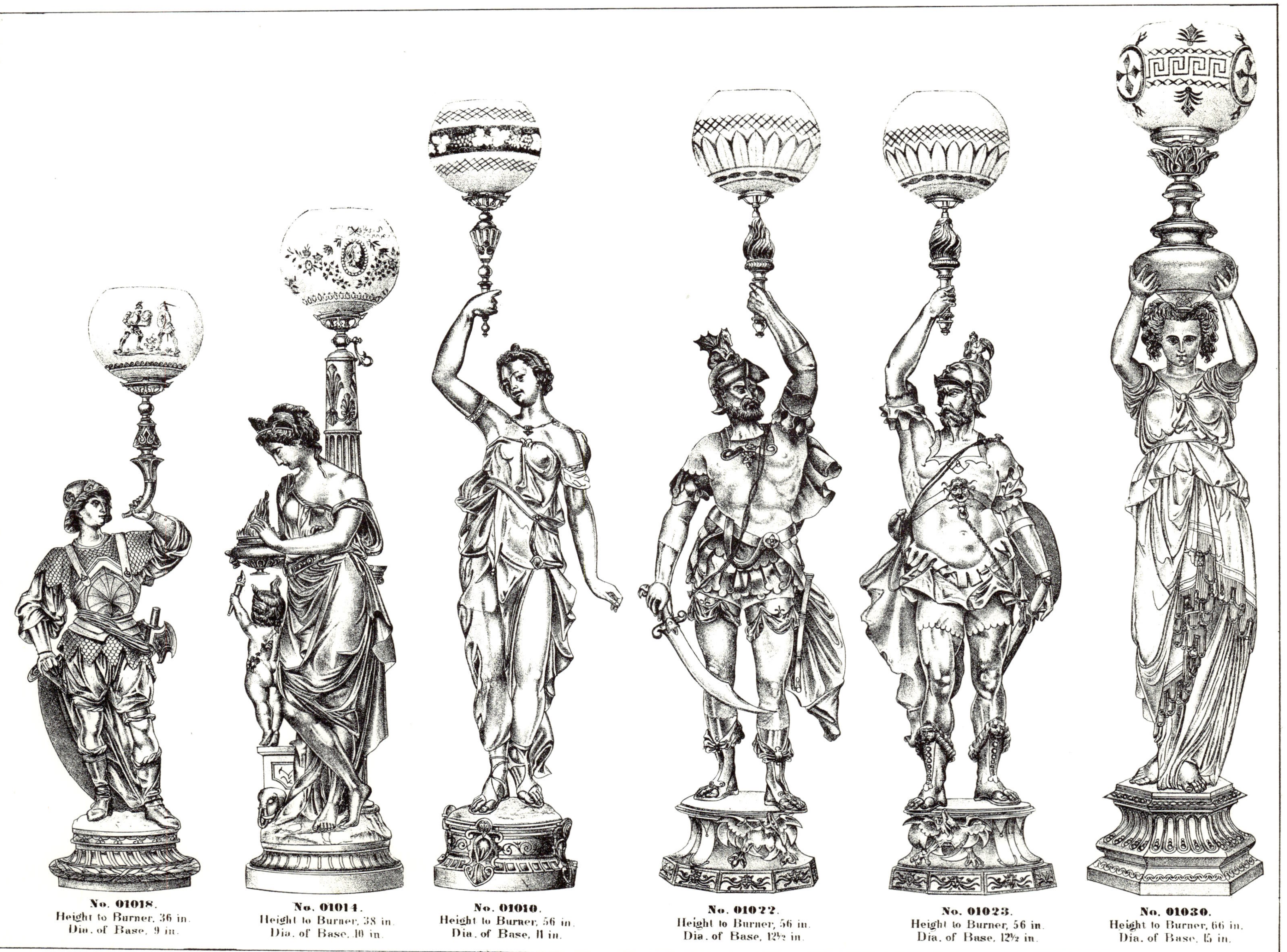

No. 01018.
Height to Burner, 36 in.
Dia. of Base, 9 in.

No. 01014.
Height to Burner, 38 in.
Dia. of Base, 10 in.

No. 01010.
Height to Burner, 56 in.
Dia. of Base, 11 in.

No. 01022.
Height to Burner, 56 in.
Dia. of Base, 12½ in.

No. 01023.
Height to Burner, 56 in.
Dia. of Base, 12½ in.

No. 01030.
Height to Burner, 66 in.
Dia. of Base, 15 in.

No. 01017.
Height to Burner, 46 in.
Dia. of Base, 11 in.

No. 01016.
Height to Burner, 46 in.
Dia. of Base, 11 in.

No. 01029.
Height to Burner, 56 in.
Dia of Base, 12½ in.

No. 01028.
Height to Burner, 56 in.
Dia. of Base, 12½ in.

No. 01012.
Height to Burner, 44 in.
Dia. of Base, 9½ in.

No. 01013.
Height to Burner, 44 in.
Dia. of Base, 9½ in.

Nº 01168.
Hight to burner, 33 in.
Dia. of base, 10½ "
Nº 01162.
Hight to burner, 33 in.
Dia. of base 8½ "
Nº 01170.
Hight to burner, 30 in.
Dia. of base. 7½ "
Nº 01174.
Hight to burner, 27 in.
Dia. of base, 7½ "
Nº 01172.
Hight to burner, 30 in.
Dia. of base. 7½ "
Nº 01176.
Hight to burner, 27 in.
Dia. of base. 7½ "
Nº 01164.
Hight to burner. 33 in.
Dia of base, 8½ "
Nº 01166.
Hight to burner 33 in.
Dia. of base. 10½ "

1 Lt. Nº 1500.
Length 15 in.
1 Lt. Nº 1512.
Length 12½ in.
6 Lt. Nº 2588.
also in 3 & 4 Lts.
Hight 46 in. Spread 27 in.
4 Lt. Nº 2616.
also in 2, 3 & 6 Lts.
Hight 42 in. Spread 26 in.
1 Lt. Nº 1506.
Length 13 in.
12 Lt. Nº 2556. also in 3, 4 & 6 Lts.
Hight 12 Lt. 60 in Hight 3, 4 & 6 Lts. 54 in
Spread 12 34 Spread 3, 4 & 6, 32
1 Lt. Nº 1510
Length 11 in

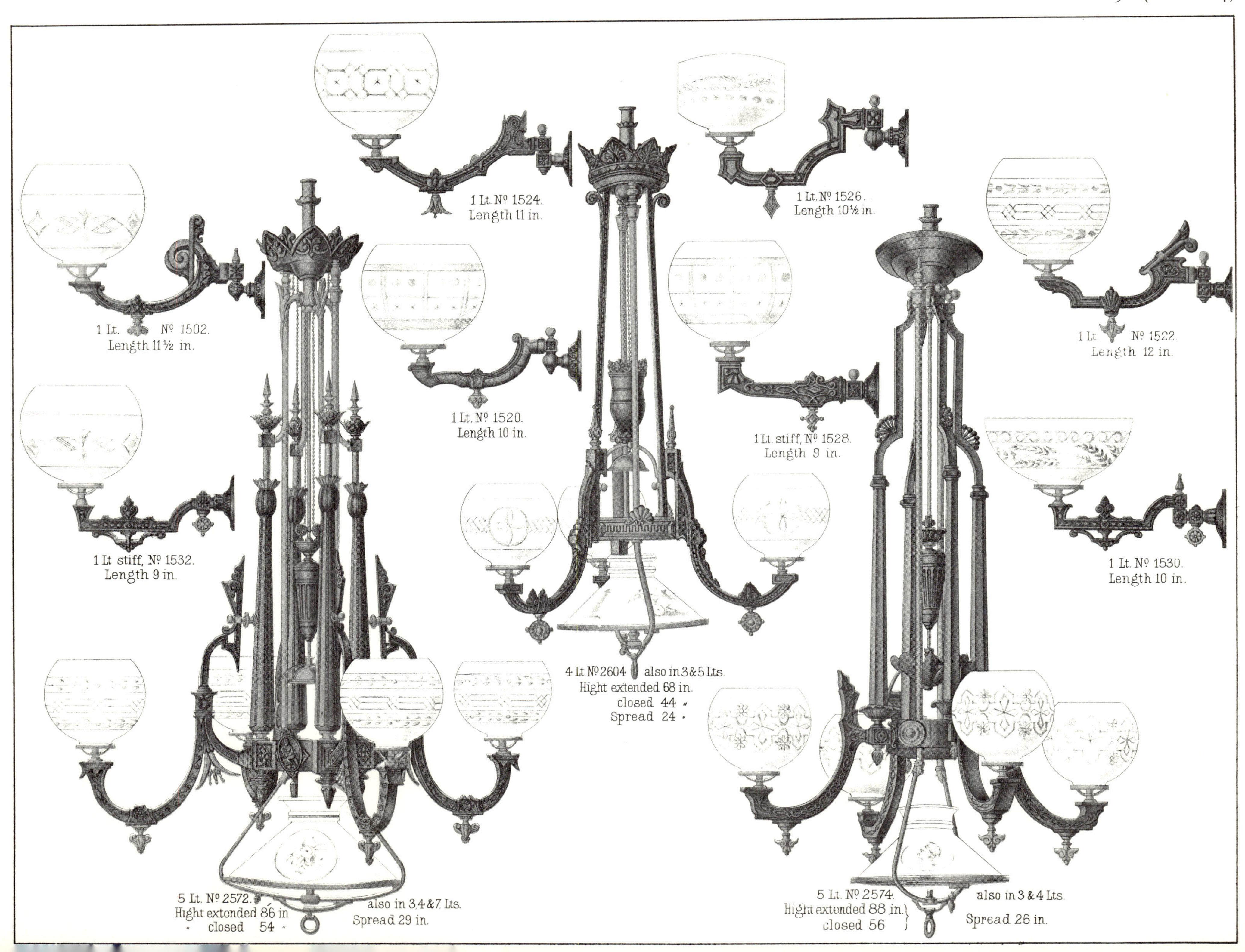
1 Lt. Nº 1524.
Length 11 in.
1 Lt. Nº 1526.
Length 10½ in.
1 Lt. Nº 1502.
Length 11½ in.
1 Lt. Nº 1522.
Length 12 in.
1 Lt. Nº 1520.
Length 10 in.
1 Lt. stiff, Nº 1528.
Length 9 in.
1 Lt stiff, Nº 1532.
Length 9 in.
1 Lt. Nº 1530.
Length 10 in.
4 Lt Nº 2604 also in 3 & 5 Lts.
Hight extended 68 in.
closed 44 "
Spread 24 "
5 Lt. Nº 2572. also in 3,4 & 7 Lts.
Hight extended 86 in
" closed 54 "
Spread 29 in.
5 Lt. Nº 2574. also in 3 & 4 Lts.
Hight extended 88 in.
closed 56
Spread 26 in.

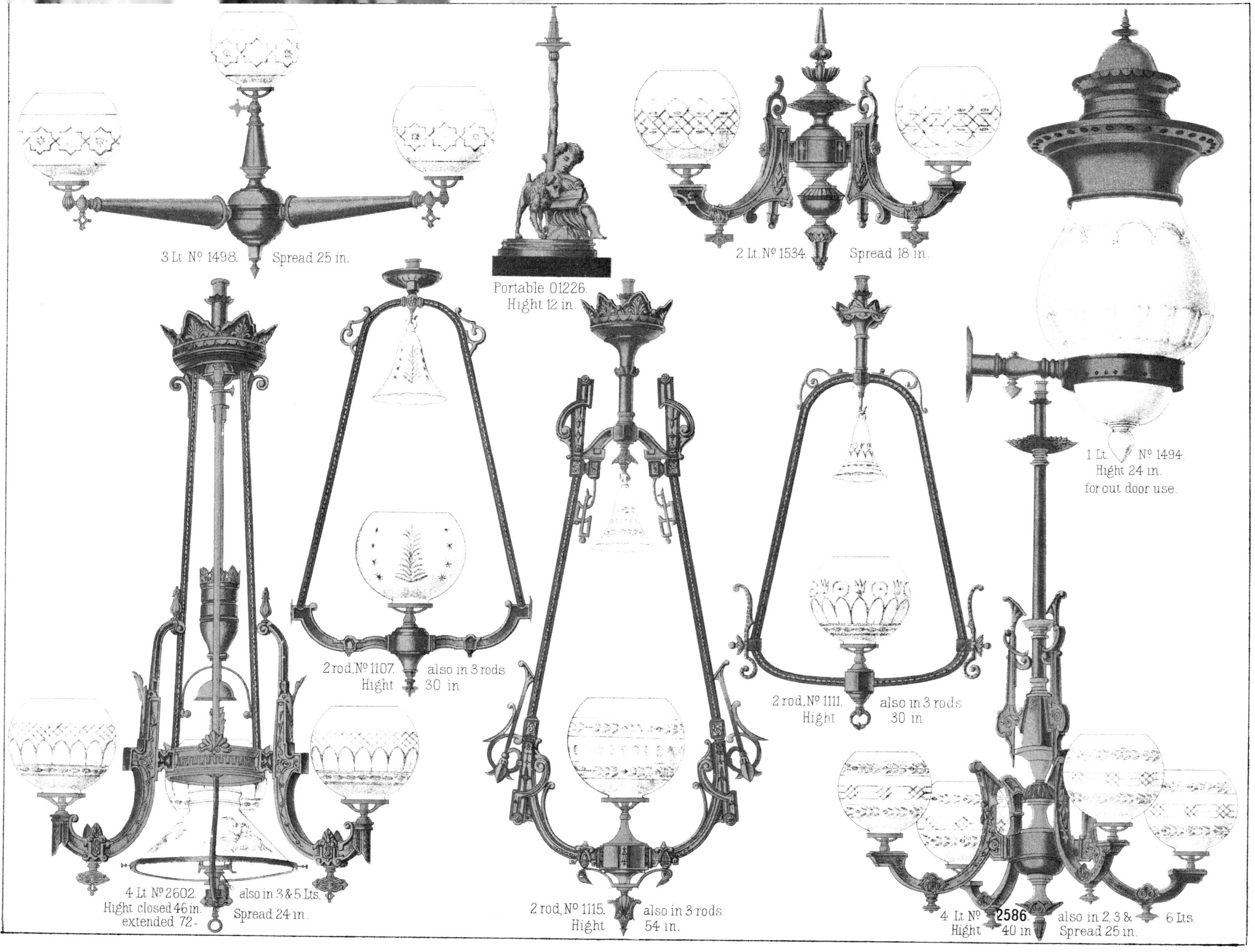
3 Lt Nº 1498. Spread 25 in.
Portable 01226. Hight 12 in.
2 Lt. Nº 1534. Spread 18 in.
1 Lt. Nº 1494. Hight 24 in. for out door use.
2 rod, Nº 1107. Hight also in 3 rods 30 in
2 rod, Nº 1111. Hight also in 3 rods 30 in
4 Lt Nº 2602. Hight closed 46 in. extended 72. also in 3 & 5 Lts. Spread 24 in.
2 rod, Nº 1115. Hight also in 3 rods 54 in.
4 Lt Nº 2586. Hight 40 in also in 2, 3 & 6 Lts. Spread 25 in.

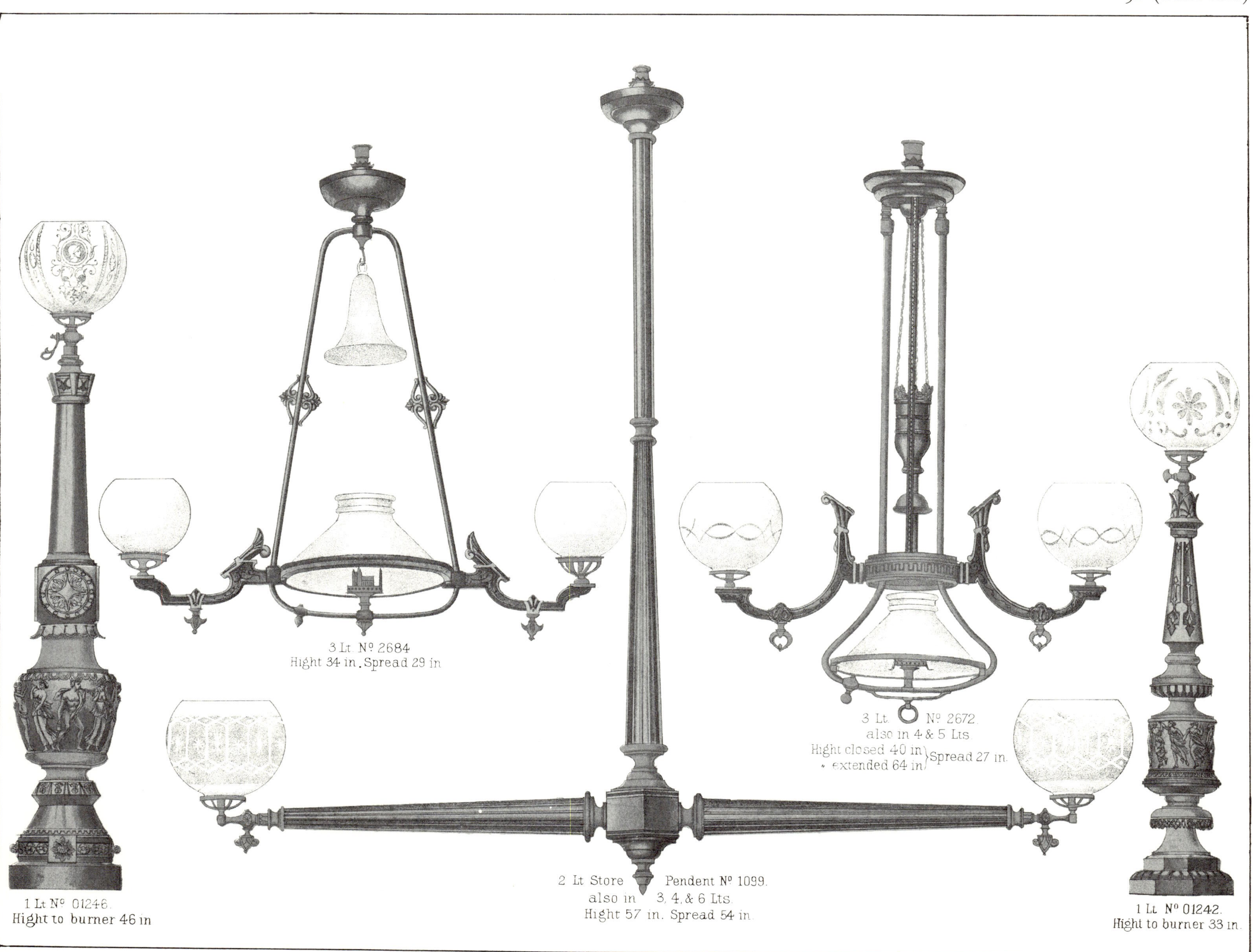
1 Lt Nº 01246.
Hight to burner 46 in
3 Lt. Nº 2684
Hight 34 in. Spread 29 in
2 Lt Store Pendent Nº 1099.
also in 3, 4, & 6 Lts.
Hight 57 in. Spread 54 in.
3 Lt. Nº 2672.
also in 4 & 5 Lts.
Hight closed 40 in
" extended 64 in
Spread 27 in.
1 Lt. Nº 01242.
Hight to burner 33 in.

1 Lt. N° 01240.
Hight to burner 26 in.
5 Lt N° 2658 also in 4 & 7 lts.
Hight closed 50 in
extended 82 in
Spread 32 in
4 Lt. N° 2668. also in 2,3,&6 lts.
Hight 45 in. Spread 28 in.
6 Lt. N° 2660 also in 3 & 4 lts.
Hight 50 in. Spread 32 in.
1 Lt. N° 01238.
Hight to burner 26 in.

1 Lt. Nº 1125.
Hight 33 in.
6 Lt. Nº 2664. also in 4 Lts
Hight 56 in. Spread 32 in.
5 Lt. Nº 2686
also in 3 & 4 Lts.
Hgt extended 78 in } Spread
" closed 48 " } 27 in.
6 Lt. Nº 2554 - also in 4 Lts.
Hight 66 in. Spread 36 in.
1 Lt. Nº 1101.
Hight 32 in.

4 Lt. Nº 2618. also in 3 & 6 Lts.
Hight 66 in. Spread 32 in.

1 Lt. Nº 1550.
Hight to burner 19 in.

4 Lt. Nº 2614. also in 3 Lts.
Hight 56 in. Spread 26 in.

Nº 1121. 3 rods.
Hight 62 in.

Nº 01228 - 1 Lt.
Hight to burner 50 in.

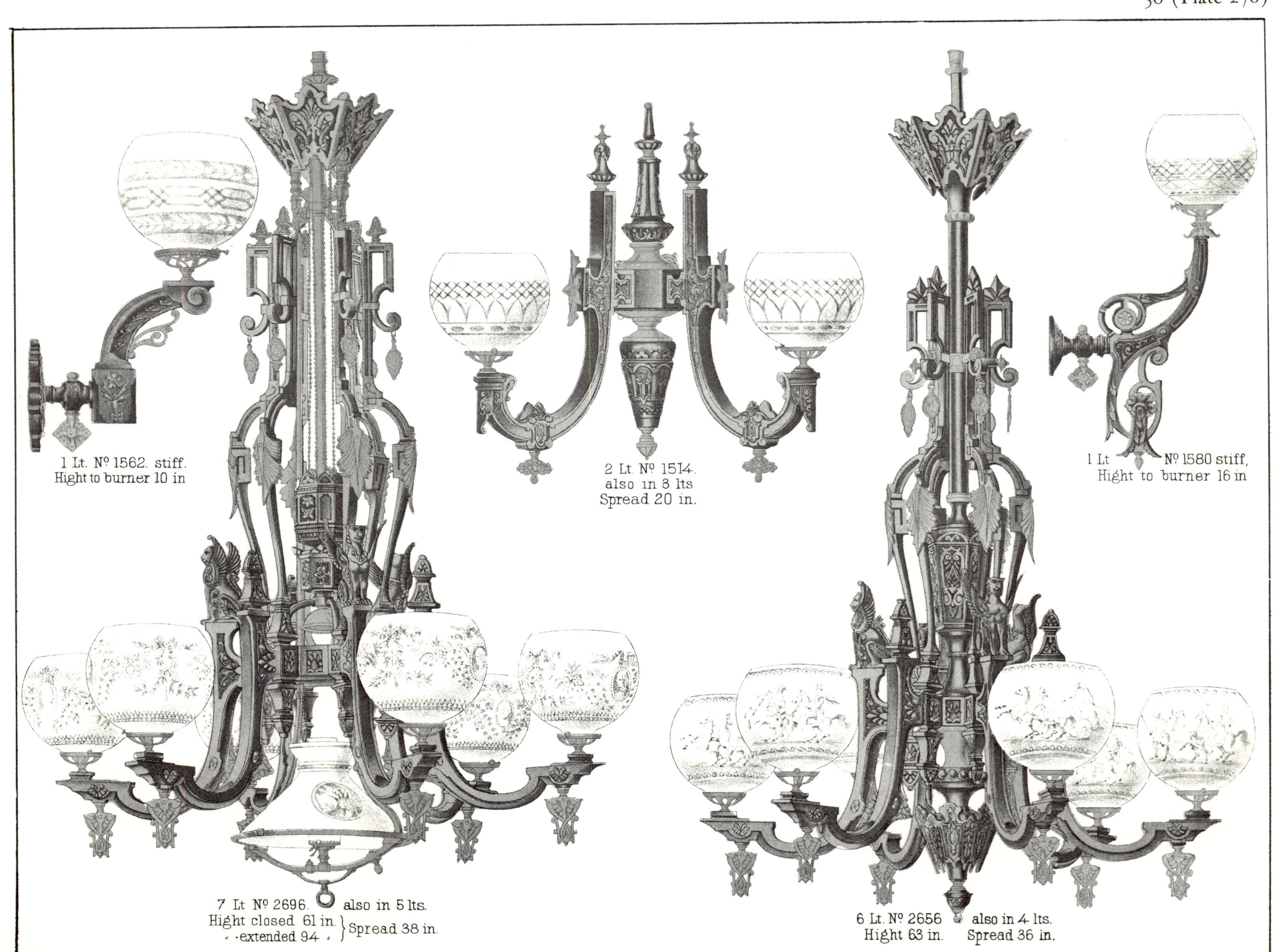
1 Lt. Nº 1562. stiff.
Hight to burner 10 in
2 Lt. Nº 1514.
also in 3 lts
Spread 20 in.
1 Lt Nº 1580 stiff,
Hight to burner 16 in
7 Lt Nº 2696. also in 5 lts.
Hight closed 61 in.
" extended 94 "
Spread 38 in.
6 Lt. Nº 2656 also in 4 lts.
Hight 63 in.
Spread 36 in.

4 Lt. Nº 01232. also in 1 & 5 lts.
Hight to burner, 4 & 5 lts 82 in.
" " " 1 " 66 "

6 Lt. Nº 2698
also in 4 lts. Hight 58 in. Spread 34 in.

4 Lt. Nº 01068. also in 1 & 5 lts.
Hight to burner, 4 & 5 lts 82 in.
" " " 1 " 66 "

2 Lt. Nº 1508.
also in 3 lts.
Spread 14 in.
2 Lt. Nº 1552.
also in 3 lts.
Spread 21 in.
3 rods, Nº 1163. also in 2 rods.
Hight 48 in.
6 Lt. Nº 2638. also in 4 lts.
Hight 62 in. Spread 42 in.
3 rods, Nº 1131. to slide.
Hight 51 in.

1 Lt. Nº 1586.
Length 13 in.
3 Lt. toilet Nº 2680.
also in 2 & 4 lts
Hight 36 in.
Spread 16 in
3 Lt. toilet, Nº 2652.
also in 2 & 4 lts.
Hight 36 in. Spread 14 in.
3 Lt Nº 2666.
also in 2, 4 & 6, lts.
Hight 43 in
Spread 26 in.
4 Lt. Nº 2700.
also in 2, 3 & 6 lts,
Hight 45 in.
Spread 2, 3 & 4 Lts. 28 in.
" 6 " 31 "
Nº 1143. 2 Lt Store Pendant
Hight 49 in
also in 4 lts
Spread 56 in.

1 Lt. Bracket. Nº 1610.
2 joints, also in 1 joint.
Length 21 in.
3 Lt. Nº 2646 also in 2 & 4 lts.
Hight 33 in. Spread 14 in.
3 Lt. Nº 2642 also in 2 & 4 lts.
Hight 34 in. Spread 13 in.
3 Lt Nº 2640. also in 2 & 4 lts.
Hight 34 in. Spread 14 in.
3 Lt. Nº 2644. also in 2 & 4 lts.
Hight 33 in. Spread 14 in.
3 Lt. Nº 2704. also in 2 & 4 lts.
Hight 60 in. Spread 22 in.
3 Lt. Nº 2682. also in 2 & 4 lts.
Hight 30 in. Spread 13 in.
3 Lt. Nº 2678. also in 2 & 4 lts.
Hight 30 in. Spread 13 in.

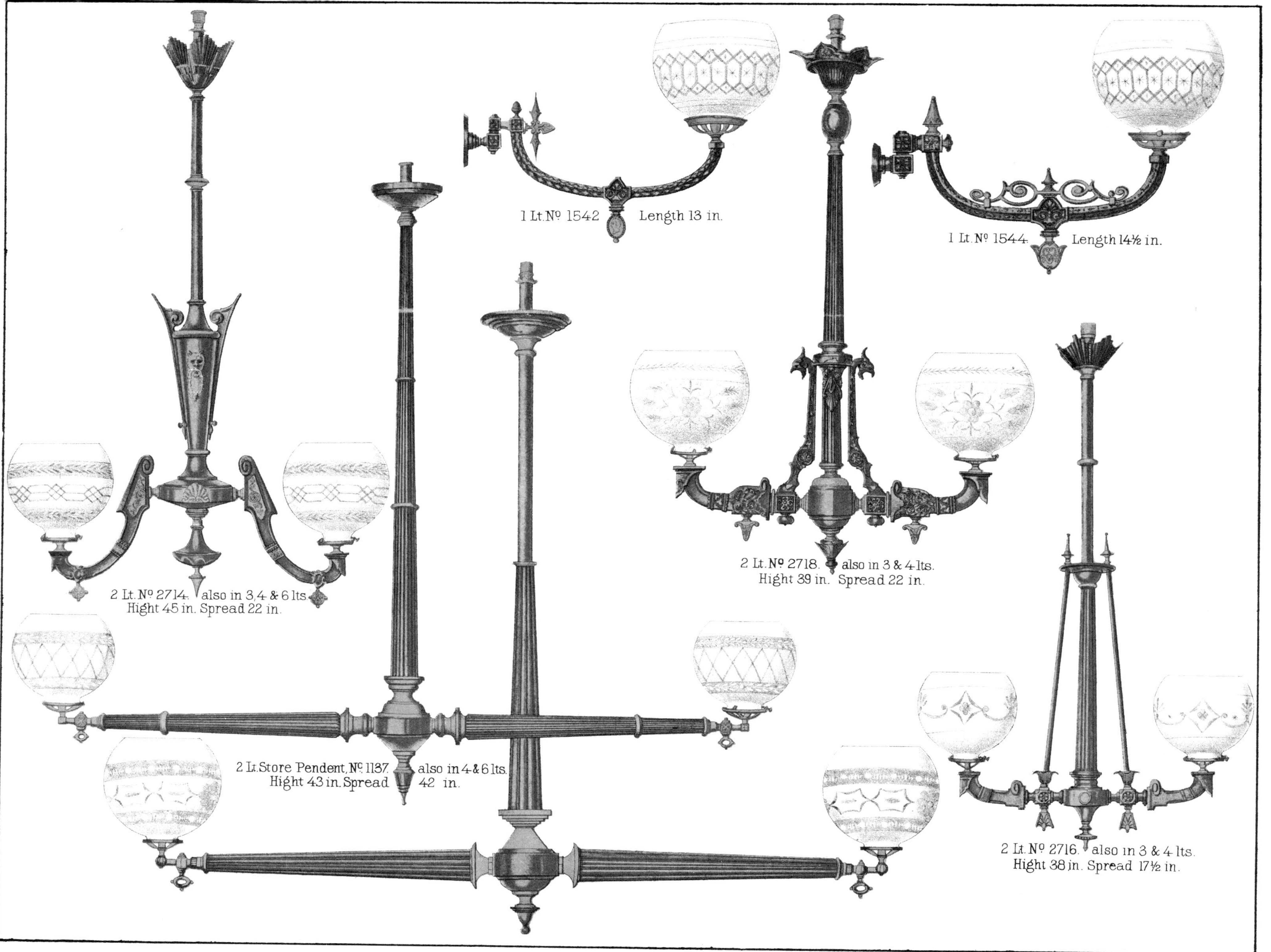
1 Lt. No 1542 Length 13 in.
1 Lt. No 1544. Length 14½ in.
2 Lt. No 2714. also in 3, 4 & 6 lts.
Hight 45 in. Spread 22 in.
2 Lt. No 2718. also in 3 & 4 lts.
Hight 39 in. Spread 22 in.
2 Lt. Store Pendent, No 1137. also in 4 & 6 lts.
Hight 43 in. Spread 42 in.
2 Lt. No 2716. also in 3 & 4 lts.
Hight 38 in. Spread 17½ in.

2 Lt. Nº 2720. also in 3, 4 & 6 lts.
Hight 36 in. Spread 2, 3 & 4 lts. 22 in.
6 " 25 "
2 Lt. Nº 2726. also in 3 & 4 lts.
Hight 34 in. Spread 19 in.
2 Lt. Nº 2724 also in 3 & 4 lts.
Hight 36 in. Spread 22 in.
2 Lt Store Pendent, Nº 1153. also in 4 & 6 lts.
Hight 44 in. Spread 37 in.
2 Lt. Store Pendent Nº 1149. also in 4 & 6 lts.
Hight 48 in. Spread 56 in.

1 Lt. Nº 1566. Length 11 in.
2 Lt. Nº 1574
Spread 18 in.
1 Lt. Nº 1504.
Length 12 in.
2 Lt. Nº 2740. also in 3 & 4 lts.
Hight 34 in Spread 21 in
2 Lt. Nº 2670.
also in 3 & 4 lts.
Hight 32 in. Spread 20 in.
2 Lt Nº 2742. also in 3, 4 & 6 lts
Hight 34 in. Spread 2, 3 & 4 lts 22 in.
" 6 " 25 "
2 Lt. Nº 2744. also in 3, 4 & 6 lts
Hight 36 in. Spread 2, 3 & 4 lts. 23 in.
6 " 26 "

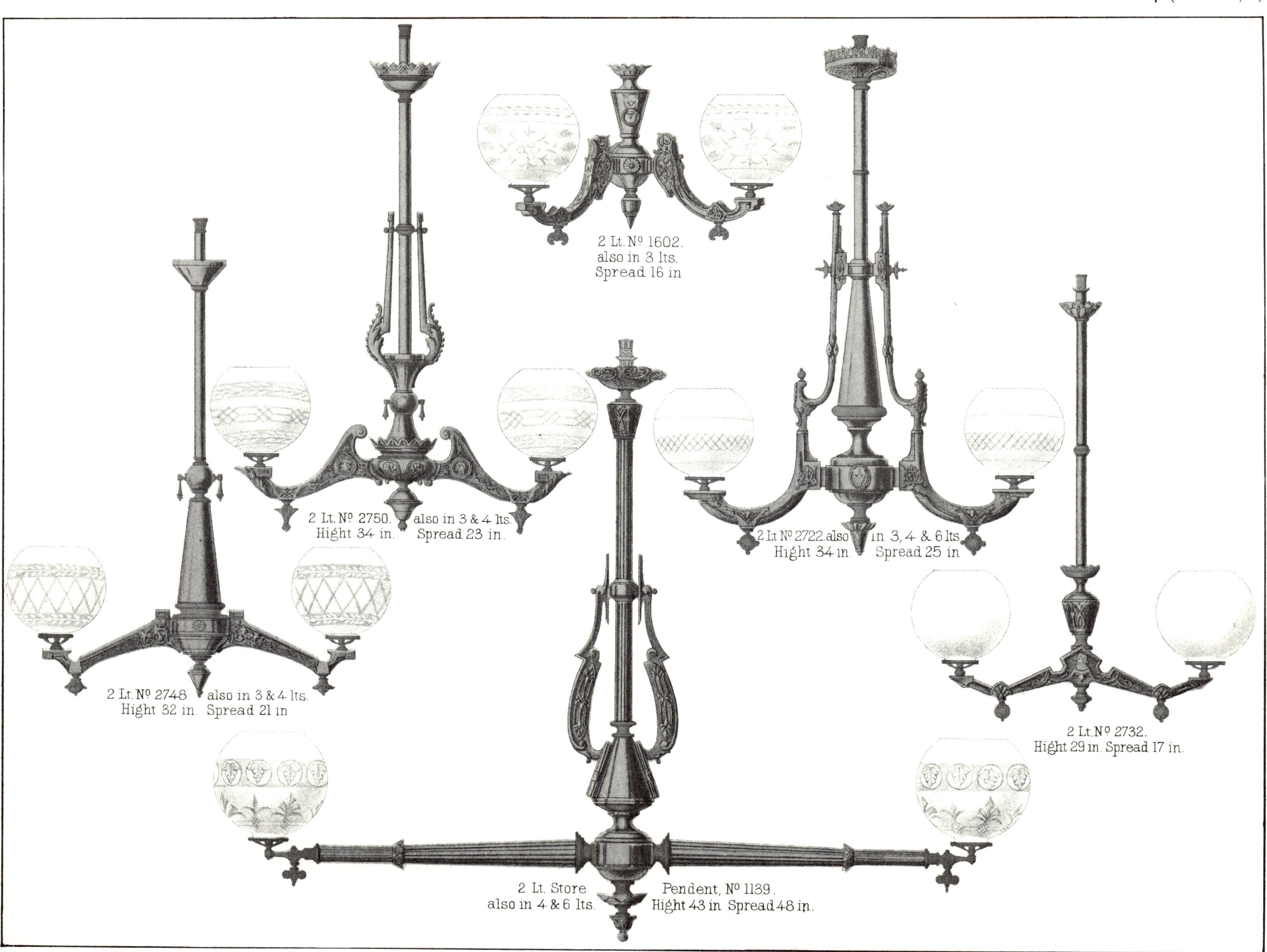
2 Lt. Nº 1602.
also in 3 lts.
Spread 16 in
2 Lt. Nº 2750. also in 3 & 4 lts.
Hight 34 in. Spread 23 in.
2 Lt Nº 2722. also in 3, 4 & 6 lts.
Hight 34 in. Spread 25 in
2 Lt. Nº 2748 also in 3 & 4 lts.
Hight 32 in. Spread 21 in
2 Lt. Nº 2732.
Hight 29 in. Spread 17 in.
2 Lt. Store Pendent, Nº 1139.
also in 4 & 6 lts. Hight 43 in Spread 48 in.

1 Lt. slide, Nº 1147.
Hight closed 43 in.
" extended 66 "

3 Rod, Nº 1159.
also in 2 rods. Hight 42 in.

3 Rod Nº 1161.
also in 2 rods, Hight 43 in

3 Rod, Nº 1155
also in 2 rods. Hight 43 in.

4 Lt. Nº 2728 also in 2,3 & 6 lts.
Hight 43 in. Spread 24 in.

4 Lt. Nº 2712. also in 2,3 & 6 lts.
Hight 50 in. Spread 28 in.

4 Lt. Billiard Pendent Nº 1029.
Hight 48 in. Spread 38 X 63 in.

4 Lt. Cluster, Nº 01254.
also in 3 lts.
Spread 19 in.

1 Lt. Nº 1576 stiff.
Hight to burner 12 in.

1 Lt. Nº 1578. stiff.
Hight to burner 11 in.

4 Lt. Cluster, Nº 01236.
also in 3 lts.
Spread 15 in.

4 Lt. Cluster Nº 01244.
also in 3 lts.
Spread 9 in.

2 Lt. Cluster, Nº 01100.
also in 3 & 4 lts.
Spread 3½ in.

3 Lt. Cluster, Nº 01294.
also in 2 lts.
Spread 4 in.

2 Lt. Nº 2752. Hight 36 in. also in 3 4 & 6 lts
Spread 2, 3 & 4 lts 25 in.
" 6 " 27 "

2 Lt. Nº 2754. Hight 37 in. also in 3 4 & 6 lts
Spread 2, 3 & 4 lts 25 in
" 6 " 27 "

2 Lt. Store Pendent, Nº 1141 also in 4 & 6 lts.
Hight 60 in. Spread 56 in.

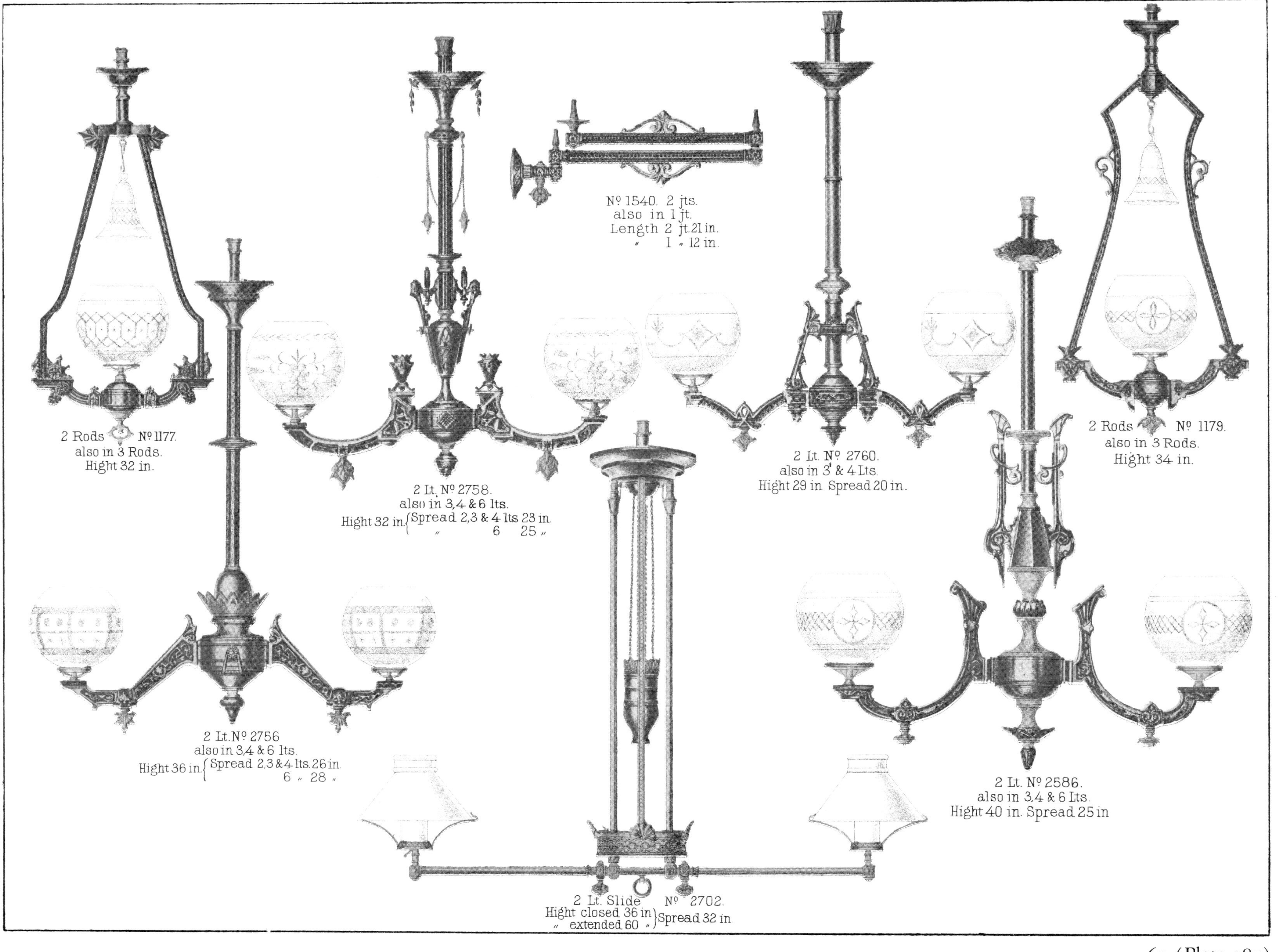

2 Rods Nº 1177.
also in 3 Rods.
Hight 32 in.
2 Lt. Nº 2758.
also in 3,4 & 6 lts.
Hight 32 in. {Spread 2,3 & 4 lts 23 in.
" 6 25 "
Nº 1540. 2 jts.
also in 1 jt.
Length 2 jt. 21 in.
" 1 " 12 in.
2 Lt. Nº 2760.
also in 3 & 4 Lts.
Hight 29 in. Spread 20 in.
2 Rods Nº 1179.
also in 3 Rods.
Hight 34 in.
2 Lt. Nº 2756
also in 3,4 & 6 lts.
Hight 36 in. {Spread 2,3 & 4 lts. 26 in.
6 " 28 "
2 Lt. Nº 2586.
also in 3,4 & 6 Lts.
Hight 40 in. Spread 25 in
2 Lt. Slide Nº 2702.
Hight closed 36 in } Spread 32 in
" extended 60 "

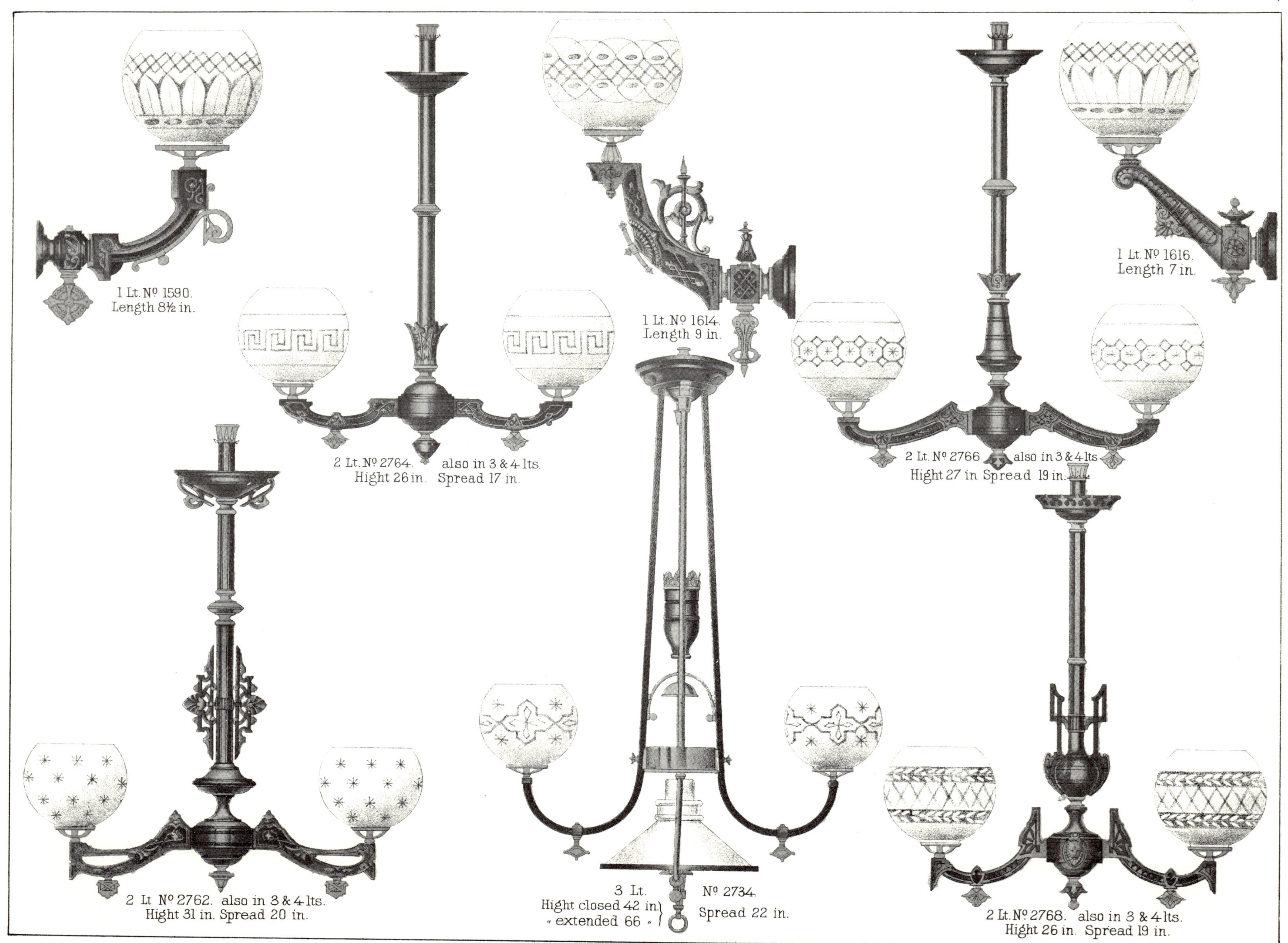
1 Lt. Nº 1590.
Length 8½ in.
2 Lt. Nº 2764. also in 3 & 4 lts.
Hight 26 in. Spread 17 in.
1 Lt. Nº 1614.
Length 9 in.
2 Lt. Nº 2766. also in 3 & 4 lts.
Hight 27 in. Spread 19 in.
1 Lt. Nº 1616.
Length 7 in.
2 Lt Nº 2762. also in 3 & 4 lts.
Hight 31 in. Spread 20 in.
3 Lt. Nº 2734.
Hight closed 42 in.
" extended 66 "
Spread 22 in.
2 Lt. Nº 2768. also in 3 & 4 lts.
Hight 26 in. Spread 19 in.

1 Lt. Nº 1584. stiff.
Length 9 in.
1 Lt. Nº 1582. stiff.
Length 8 in.
4 Lt. Nº 2770.
Hight 44 in. Spread 22 in.
4 Lt. Nº 2694.
also in 3 & 6 lts.
Hight 51 in. Spread 25 in.
1 Lt. Nº 1572.
Length 12 in.
12 Lt. Nº 2676.
Hight 54 in. Spread 36 in.
1 Lt. Nº 1588.
Length 11 in.

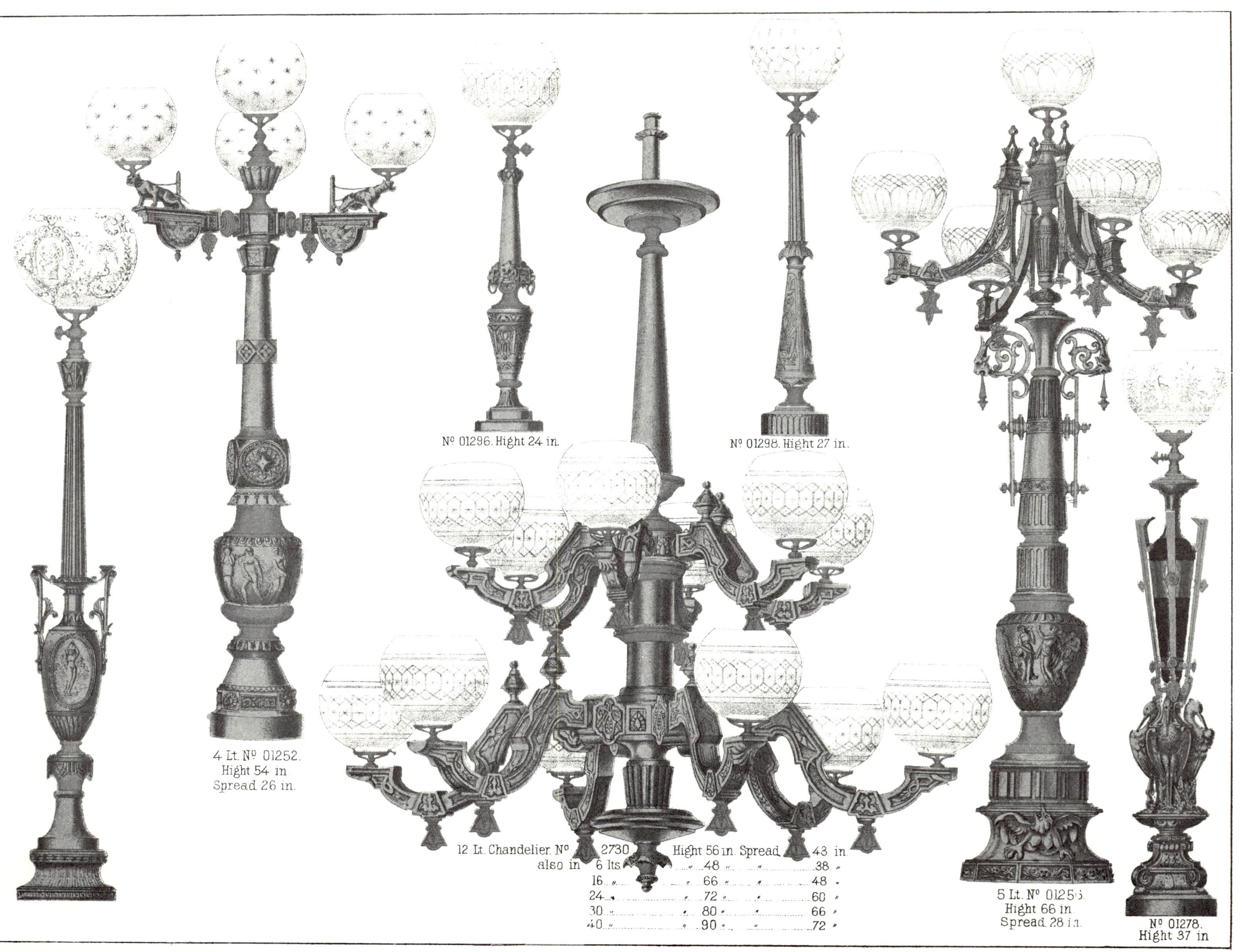

4 Lt. Nº 01252. Hight 54 in. Spread 26 in.
Nº 01296. Hight 24 in.
Nº 01298. Hight 27 in.
12 Lt. Chandelier. Nº 2730 Hight 56 in. Spread 43 in.
also in 6 lts " 48 " " 38 "
16 " " 66 " " 48 "
24 " " 72 " " 60 "
30 " " 80 " " 66 "
40 " " 90 " " 72 "
5 Lt. Nº 01255. Hight 66 in. Spread 28 in.
Nº 01278. Hight 37 in.

Portable, Nº 01300.
Hight 17 in.
Portable, Nº 01302.
Hight 15 in.
Portable, Nº 01304.
Hight 15 in.
Portable, Nº 01306.
Hight 15 in.
5 Lt. Centre lt. slide Nº 2710.
also in 4 & 7 lts.
Hight closed 48 in.
" extended 77 " } Spread 30 in.
7 Lt. Centre lt. slide Nº 2774.
also in 5 lts.
Hight closed 62 in
" extended 96 " } Spread 33 in
Portable, Nº 0734.
Hight 14 in.
2 Lt. Store Pendant, Nº 1167.
Hight 58 in.
also in 4 & 6 lts.
Spread 52 in.
Portable, Nº 01262.
Hight 14 in.

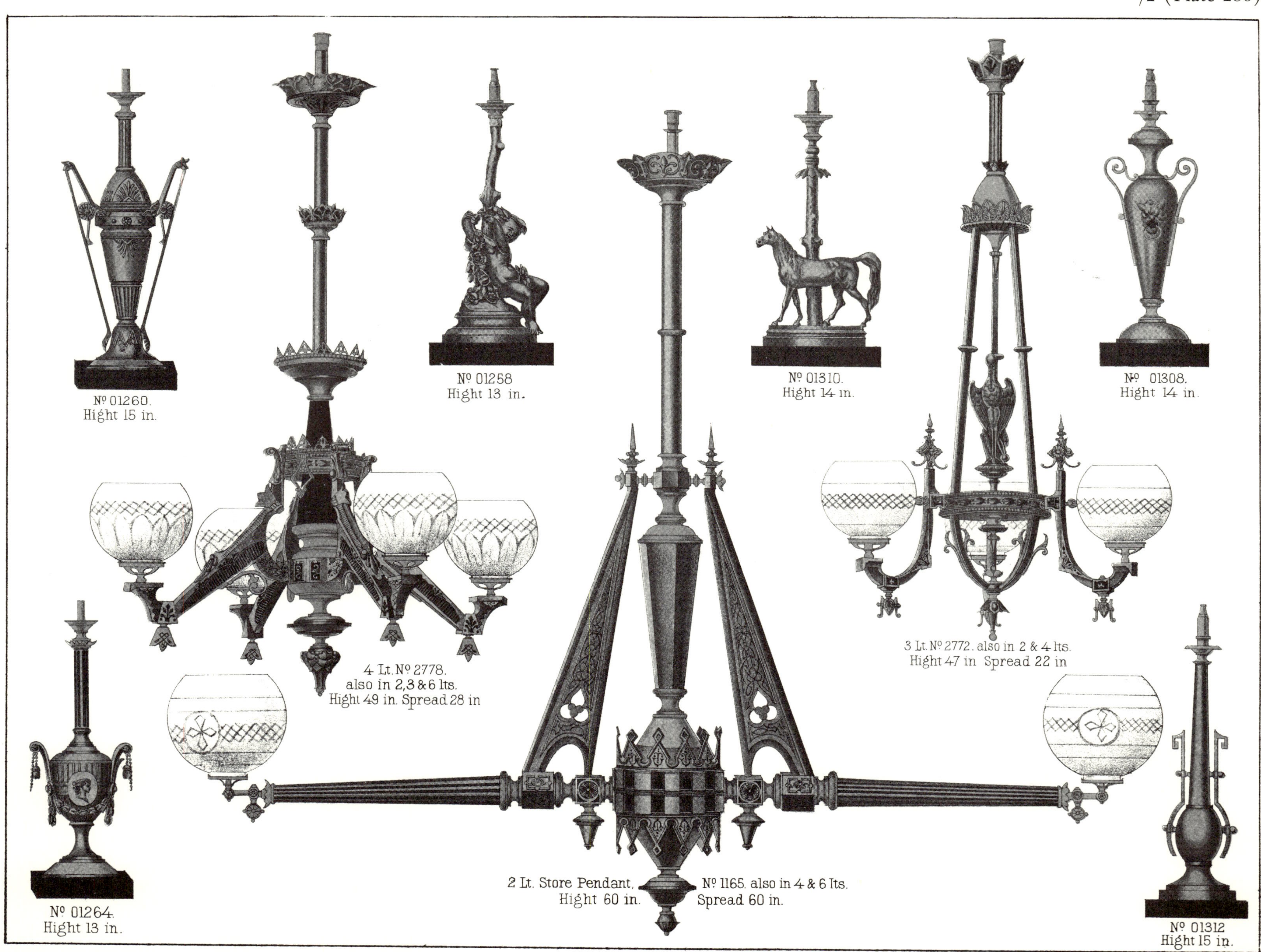

Nº 01260.
Hight 15 in.
Nº 01258
Hight 13 in.
Nº 01310.
Hight 14 in.
Nº 01308.
Hight 14 in.
4 Lt. Nº 2778.
also in 2,3 & 6 lts.
Hight 49 in. Spread 28 in
3 Lt. Nº 2772. also in 2 & 4 lts.
Hight 47 in Spread 22 in
Nº 01264.
Hight 13 in.
2 Lt. Store Pendant,
Hight 60 in.
Nº 1165. also in 4 & 6 lts.
Spread 60 in.
Nº 01312
Hight 15 in.

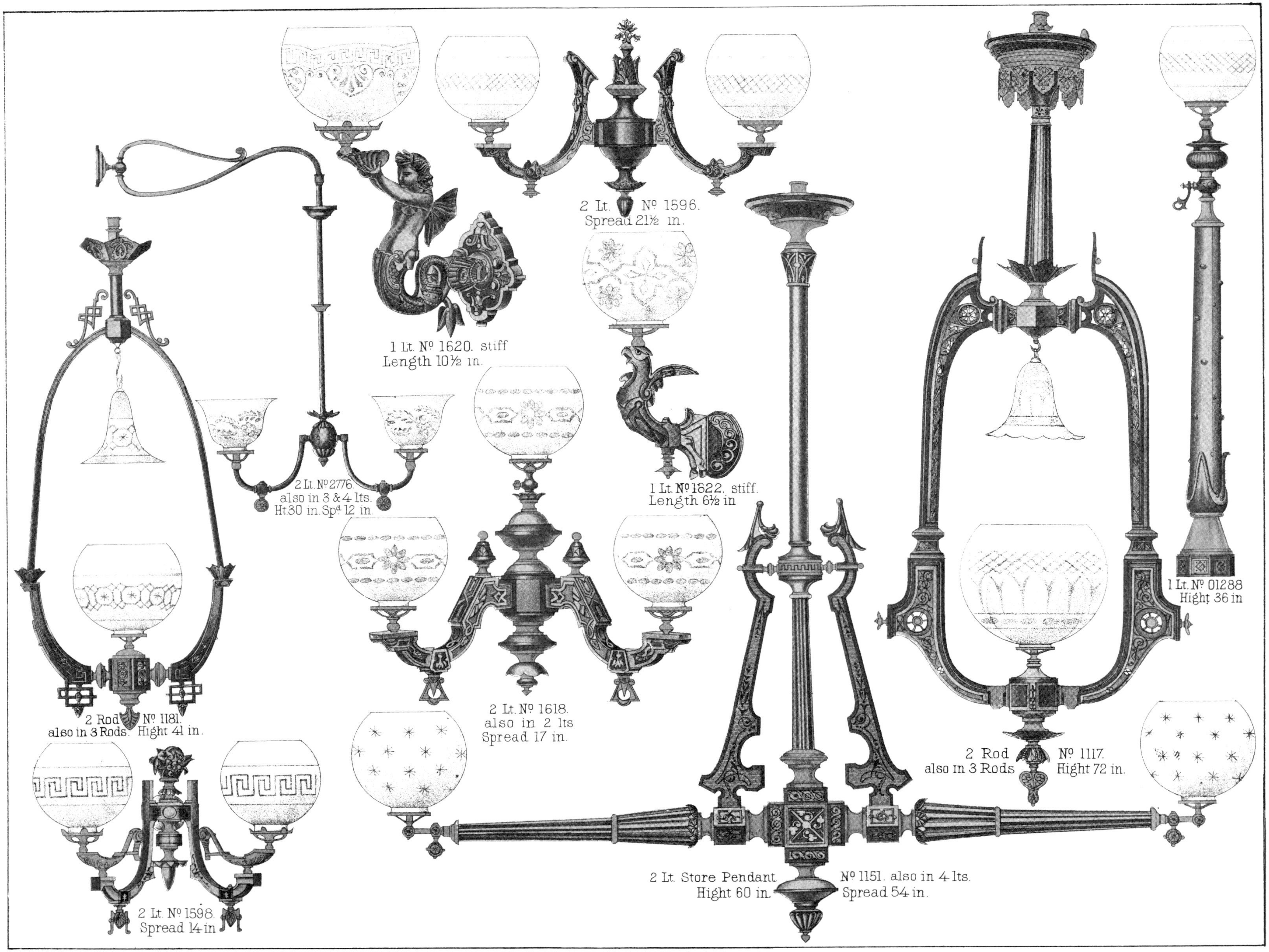

2 Lt. Nº 1596.
Spread 21½ in.

1 Lt. Nº 1620. stiff
Length 10½ in.

1 Lt. Nº 1622. stiff.
Length 6½ in

1 Lt. Nº 01288
Hight 36 in

2 Lt. Nº 2776.
also in 3 & 4 lts.
Ht 30 in. Spd 12 in.

2 Lt. Nº 1618.
also in 2 lts
Spread 17 in.

2 Rod Nº 1181.
also in 3 Rods. Hight 41 in.

2 Rod Nº 1117.
also in 3 Rods Hight 72 in.

2 Lt. Nº 1598.
Spread 14 in

2 Lt. Store Pendant. Nº 1151. also in 4 lts.
Hight 60 in. Spread 54 in.

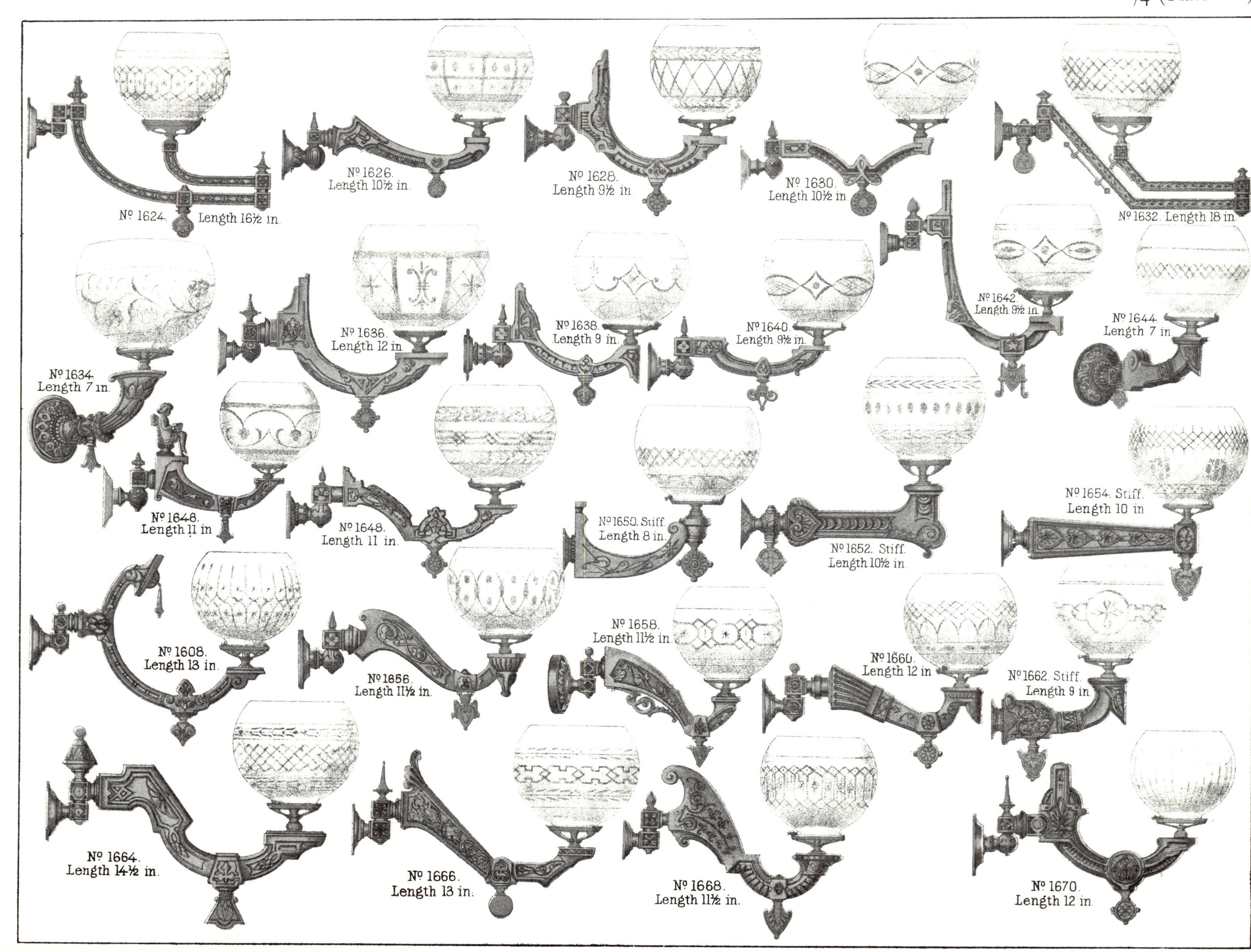

No. 1624. Length 16½ in.
No. 1626. Length 10½ in.
No. 1628. Length 9½ in.
No. 1630. Length 10½ in.
No. 1632. Length 18 in.
No. 1634. Length 7 in.
No. 1636. Length 12 in.
No. 1638. Length 9 in.
No. 1640. Length 9½ in.
No. 1642. Length 9½ in.
No. 1644. Length 7 in.
No. 1646. Length 11 in.
No. 1648. Length 11 in.
No. 1650. Stiff. Length 8 in.
No. 1652. Stiff. Length 10½ in.
No. 1654. Stiff. Length 10 in.
No. 1608. Length 13 in.
No. 1656. Length 11½ in.
No. 1658. Length 11½ in.
No. 1660. Length 12 in.
No. 1662. Stiff. Length 9 in.
No. 1664. Length 14½ in.
No. 1666. Length 13 in.
No. 1668. Length 11½ in.
No. 1670. Length 12 in.

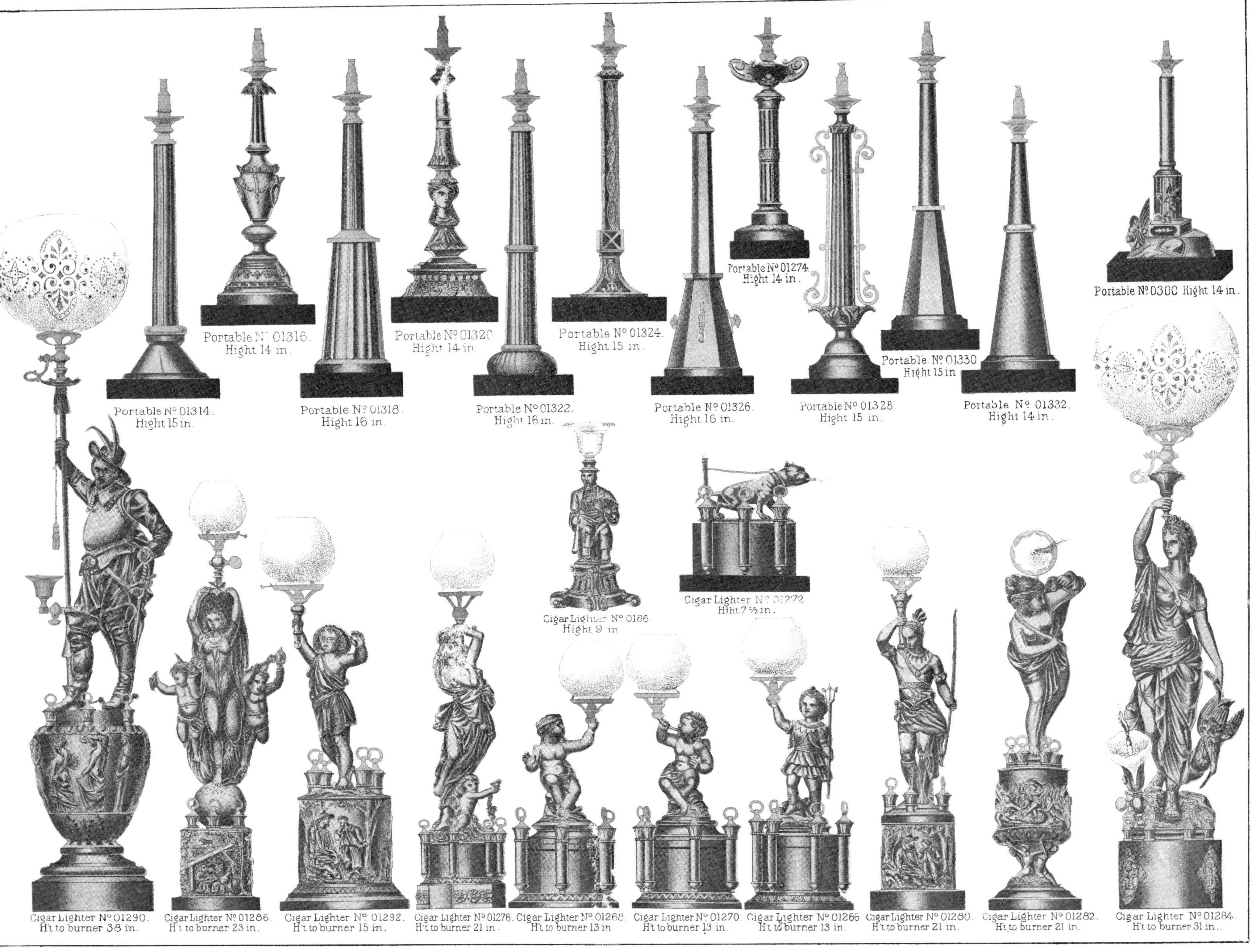
Portable Nº 01314.
Hight 15 in.
Portable Nº 01316.
Hight 14 in.
Portable Nº 01318.
Hight 16 in.
Portable Nº 01320
Hight 14 in.
Portable Nº 01322.
Hight 16 in.
Portable Nº 01324.
Hight 15 in.
Portable Nº 01326.
Hight 16 in.
Portable Nº 01274.
Hight 14 in.
Portable Nº 01328.
Hight 15 in.
Portable Nº 01330
Hight 15 in
Portable Nº 01332.
Hight 14 in.
Portable Nº 0300 Hight 14 in.
Cigar Lighter Nº 0186
Hight 9 in
Cigar Lighter Nº 01272
Hiht 7½ in.
Cigar Lighter Nº 01290.
Ht. to burner 38 in.
Cigar Lighter Nº 01286.
Ht. to burner 23 in.
Cigar Lighter Nº 01292.
Ht. to burner 15 in.
Cigar Lighter Nº 01276.
Ht. to burner 21 in.
Cigar Lighter Nº 01268.
Ht. to burner 13 in
Cigar Lighter Nº 01270.
Ht. to burner 13 in.
Cigar Lighter Nº 01266
Ht. to burner 13 in.
Cigar Lighter Nº 01280.
Ht. to burner 21 in.
Cigar Lighter Nº 01282.
Ht. to burner 21 in.
Cigar Lighter Nº 01284.
Ht. to burner 31 in.

1 jt Swing Bracket
Nº 1674. also in 2 & 3 jts.
Length 1 jt. 12 in.
2 Lt. Nº 2784. also in 3 & 4 lts.
Hight 36 in. Spread 21 in.
2 Lt. Nº 1592.
Spread 19 in.
Extension Chandelier for low ceilings.
2 Lt. Nº 2790.
Hight closed 13 in
" extended 44 " Spread 22 in
2 jt. Swing Bracket.
Nº 1674. also in 1 & 3 jts.
Length 2 jt. 21 in.
2 Lt. Nº 2786. also in 3 & 4 lts
Hight 30 in. Spread 20 in.
3 Lt. Nº 2788. also in 2 & 4 lts.
Hight 44 in. Spread 25 in.
3 Lt. Nº 2782. also in 2 & 4 lts.
Hight 48 in. Spread 26 in.
1 Lt. stiff. Nº 1560.
Length 6½ in.
2 Lt. Nº 2792.
Hight closed 34 in
" extended 56 " Spread 22 in
2 Lt. Nº 2794.
Hight closed 34 in.
" extended 56 " Spread 19 in
Match Box. Nº 5211
Hight 3½ in.

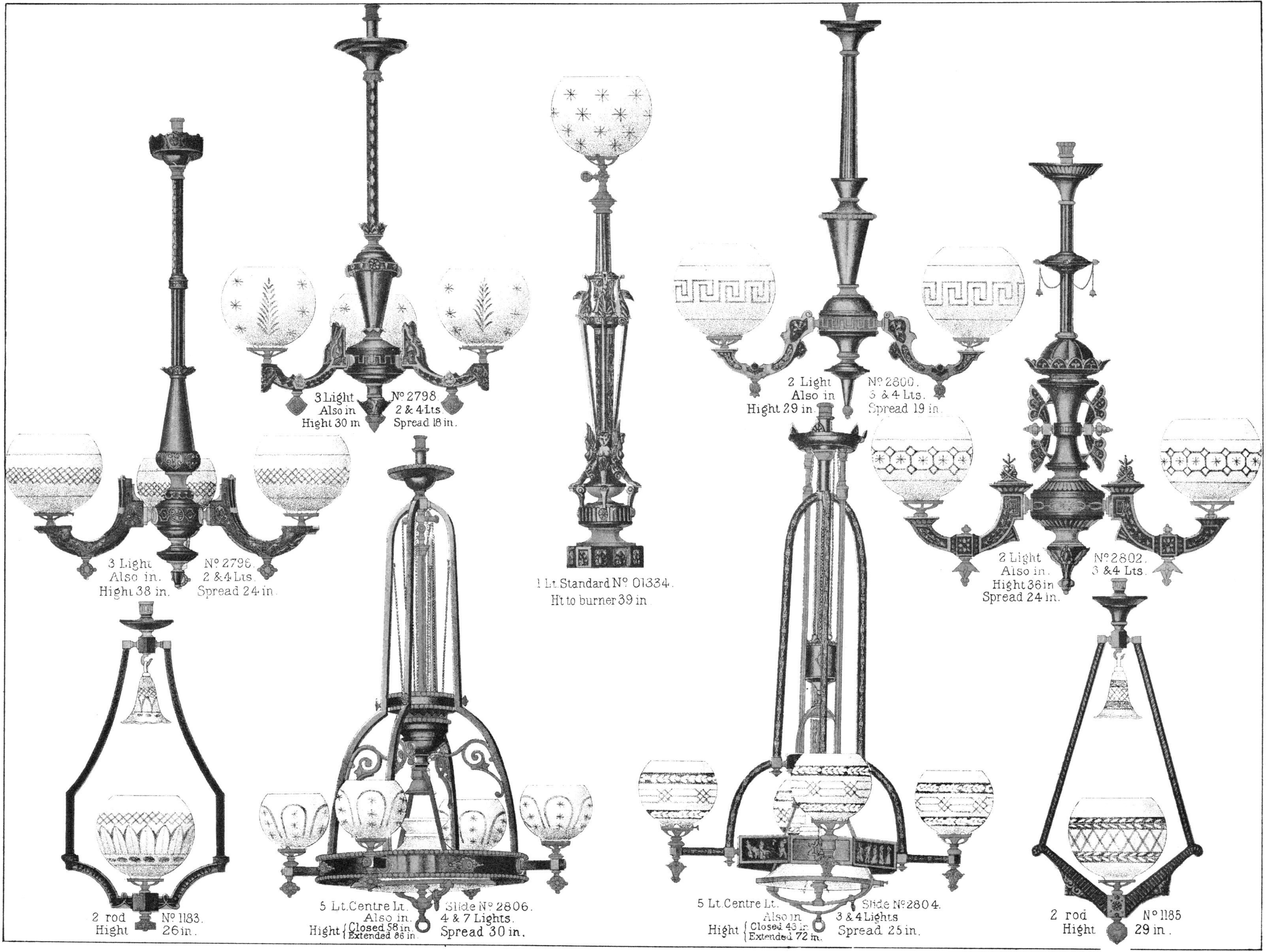
3 Light
Also in
Hight 30 in
No 2798.
2 & 4 Lts.
Spread 18 in.
3 Light
Also in.
Hight 38 in.
No 2796.
2 & 4 Lts.
Spread 24 in.
1 Lt. Standard No 01334.
Ht to burner 39 in
2 Light
Also in
Hight 29 in.
No 2800.
3 & 4 Lts.
Spread 19 in.
2 Light
Also in.
Hight 36 in
Spread 24 in.
No 2802.
3 & 4 Lts.
2 rod
Hight
No 1183.
26 in.
5 Lt. Centre Lt.
Also in.
Hight {Closed 58 in. Extended 86 in.
Slide No 2806.
4 & 7 Lights.
Spread 30 in.
5 Lt. Centre Lt.
Also in
Hight {Closed 43 in Extended 72 in.
Slide No 2804.
3 & 4 Lights
Spread 25 in.
2 rod
Hight
No 1185
29 in.

150.
392.
448.
400.
300
410
412.
414.
416.
418.
420.
422.
424.
426
428.
430.
432.
434.
436.
438.
440.
442.
444.
446.
6809
6810
6809½.
6811½.
6796.
6795.
6798.
6799.
6806.
6807
6808.
6814.

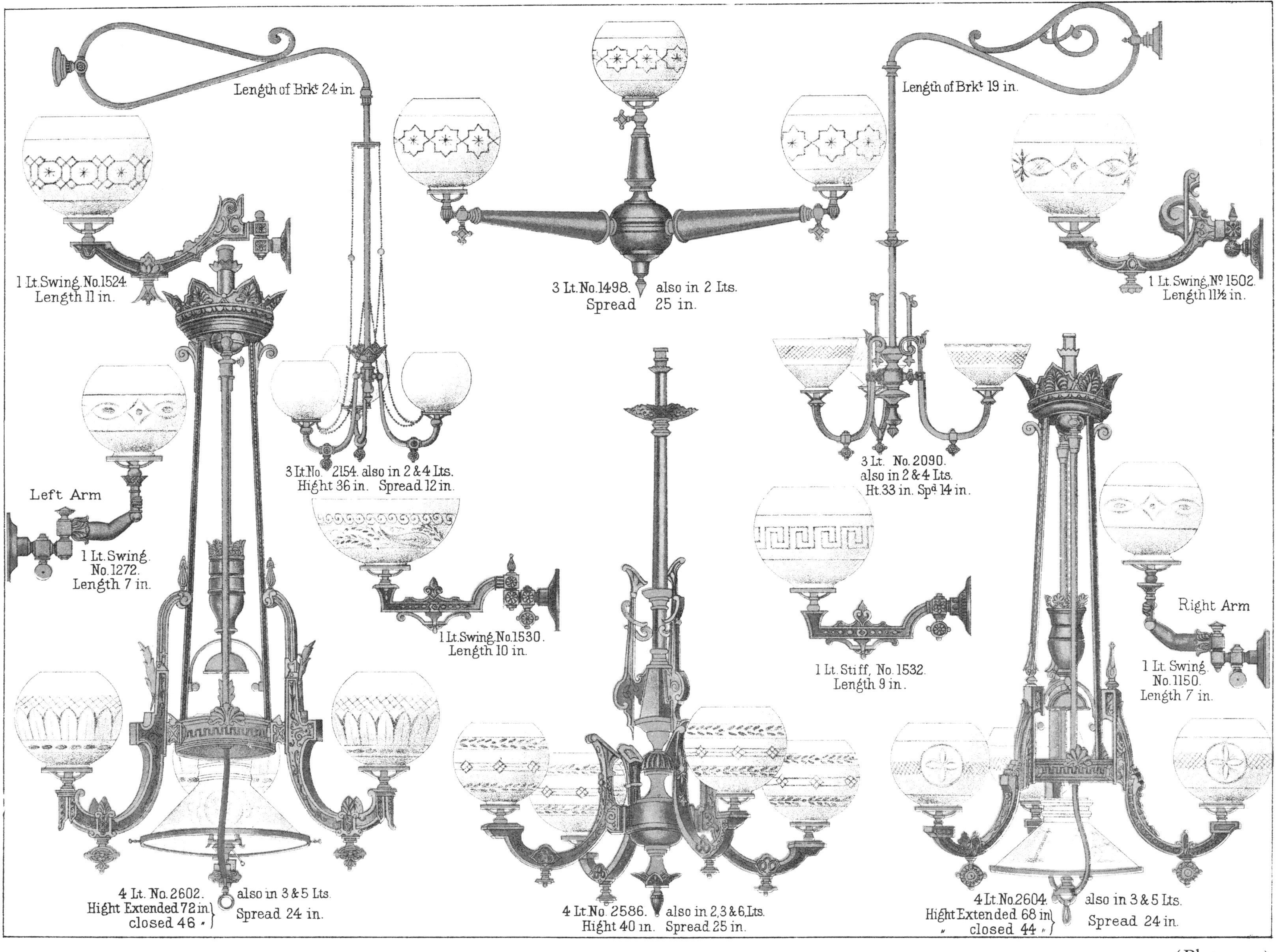
Length of Brk^t 24 in.
Length of Brk^t 19 in.
1 Lt. Swing. No. 1524.
Length 11 in.
3 Lt. No. 1498. also in 2 Lts.
Spread 25 in.
1 Lt. Swing. No 1502.
Length 11½ in.
3 Lt. No. 2154. also in 2 & 4 Lts.
Hight 36 in. Spread 12 in.
3 Lt. No. 2090.
also in 2 & 4 Lts.
Ht. 33 in. Sp^d 14 in.
Left Arm
1 Lt. Swing.
No. 1272.
Length 7 in.
1 Lt. Swing. No. 1530.
Length 10 in.
1 Lt. Stiff, No. 1532.
Length 9 in.
Right Arm
1 Lt. Swing.
No. 1150.
Length 7 in.
4 Lt. No. 2602. also in 3 & 5 Lts.
Hight Extended 72 in. closed 46 " Spread 24 in.
4 Lt. No. 2586. also in 2, 3 & 6 Lts.
Hight 40 in. Spread 25 in.
4 Lt. No. 2604. also in 3 & 5 Lts.
Hight Extended 68 in. closed 44 " Spread 24 in.

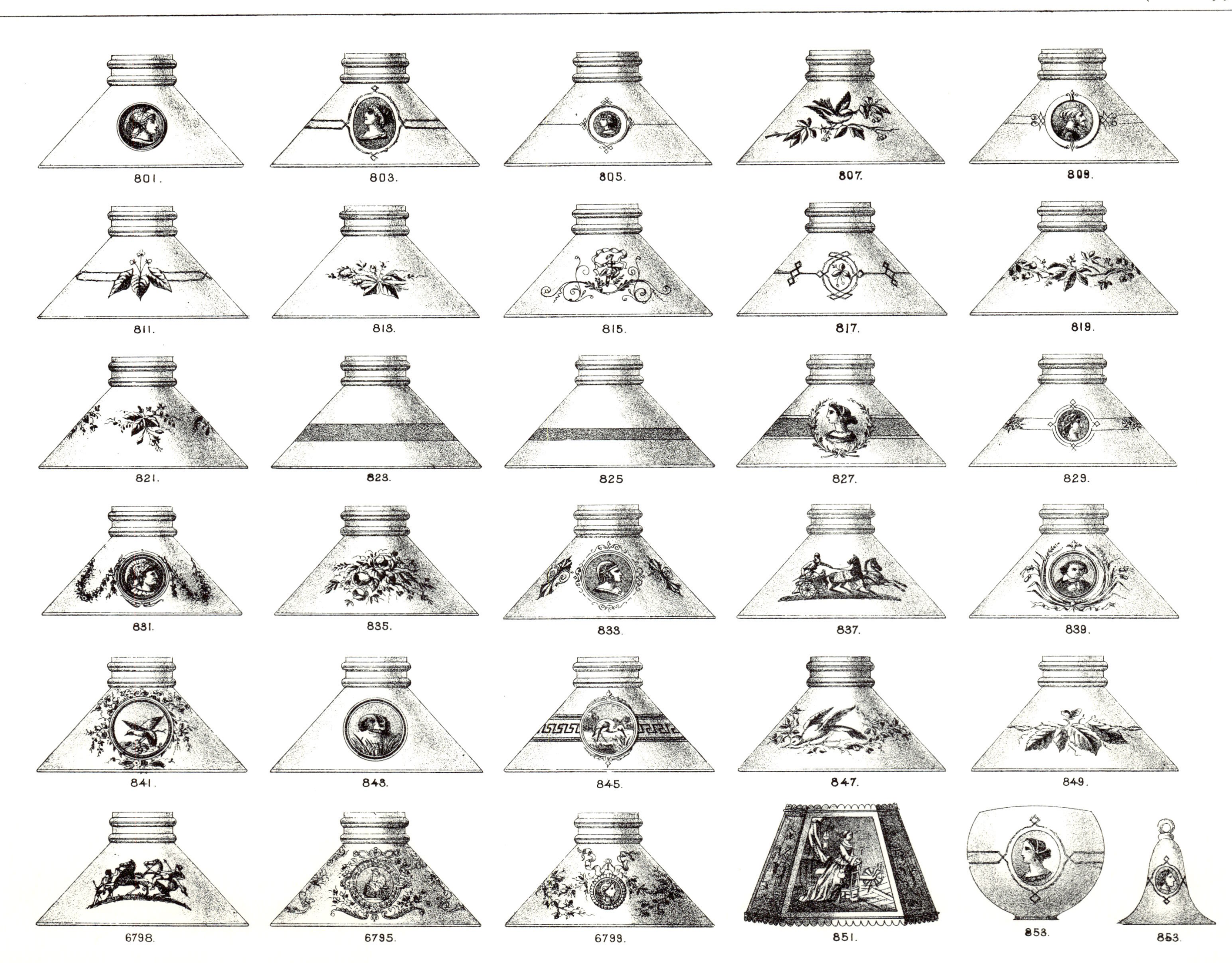
801.
803.
805.
807.
809.
811.
813.
815.
817.
819.
821.
823.
825
827.
829.
831.
835.
833.
837.
839.
841.
843.
845.
847.
849.
6798.
6795.
6799.
851.
853.
853.

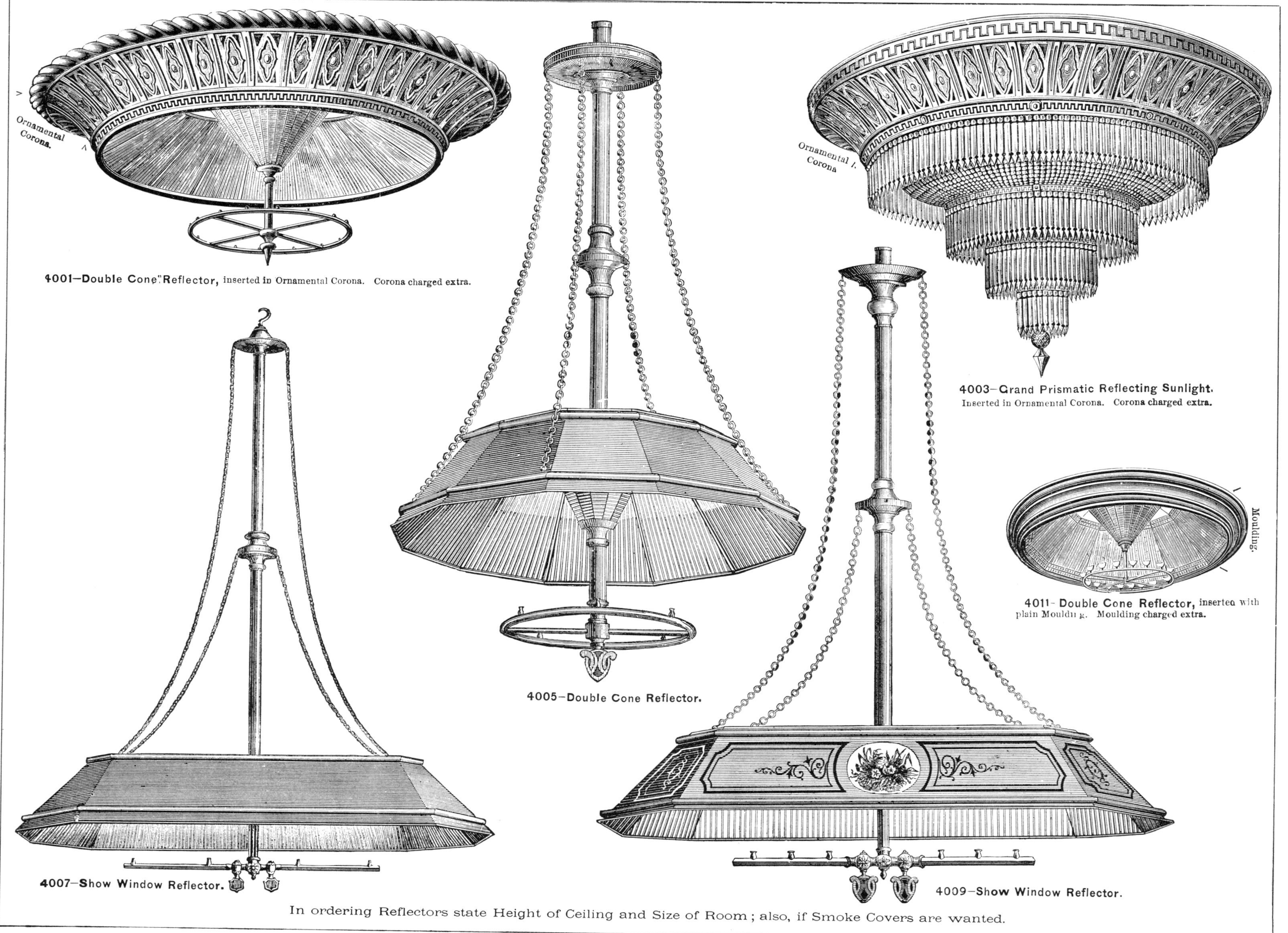

4001–Double Cone"Reflector, inserted in Ornamental Corona. Corona charged extra.

4003–Grand Prismatic Reflecting Sunlight. Inserted in Ornamental Corona. Corona charged extra.

4011–Double Cone Reflector, inserted with plain Moulding. Moulding charged extra.

4005–Double Cone Reflector.

4007–Show Window Reflector.

4009–Show Window Reflector.

In ordering Reflectors state Height of Ceiling and Size of Room; also, if Smoke Covers are wanted.

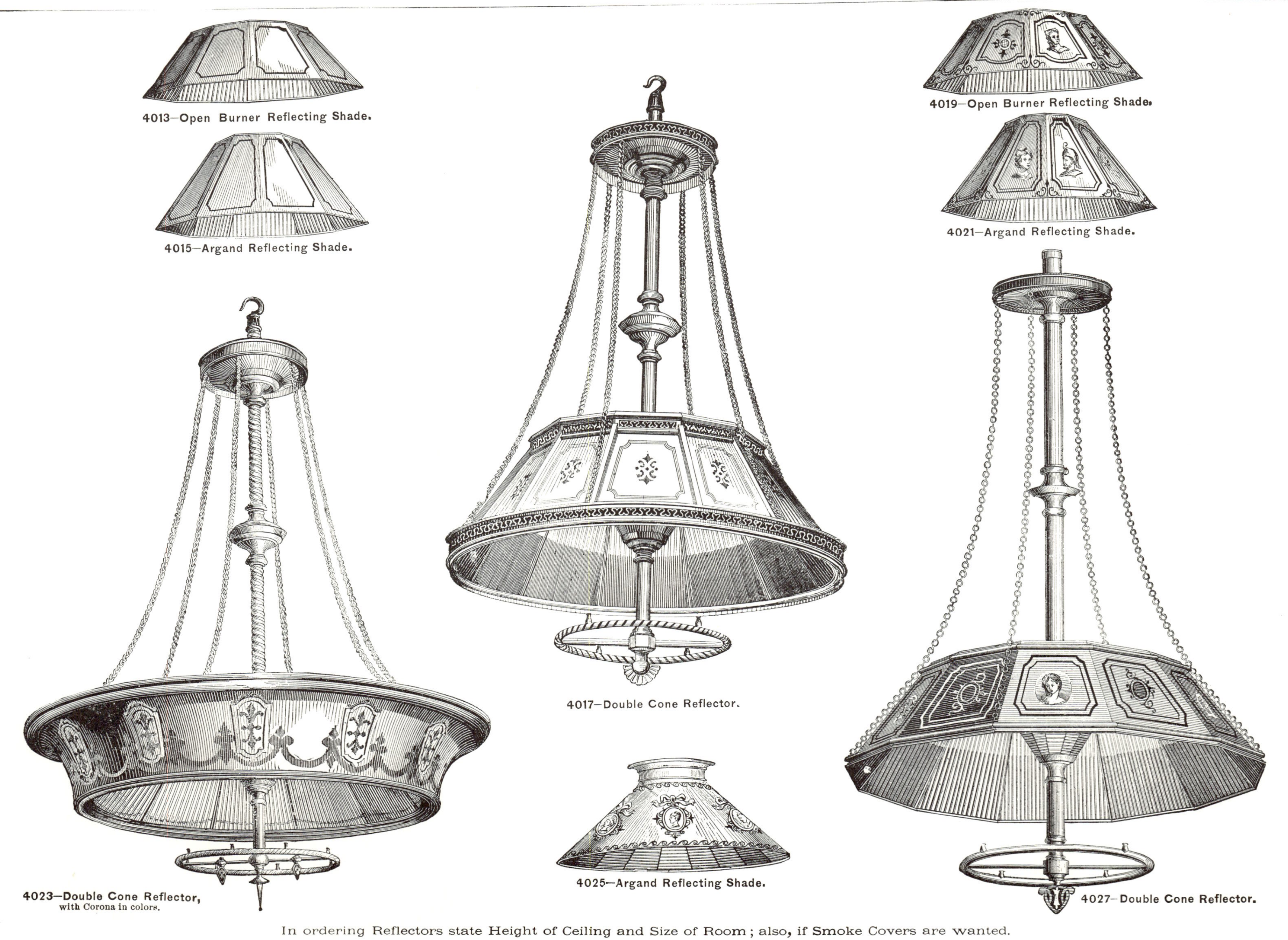

4013—Open Burner Reflecting Shade.

4015—Argand Reflecting Shade.

4019—Open Burner Reflecting Shade.

4021—Argand Reflecting Shade.

4023—Double Cone Reflector, with Corona in colors.

4017—Double Cone Reflector.

4025—Argand Reflecting Shade.

4027—Double Cone Reflector.

In ordering Reflectors state Height of Ceiling and Size of Room; also, if Smoke Covers are wanted.

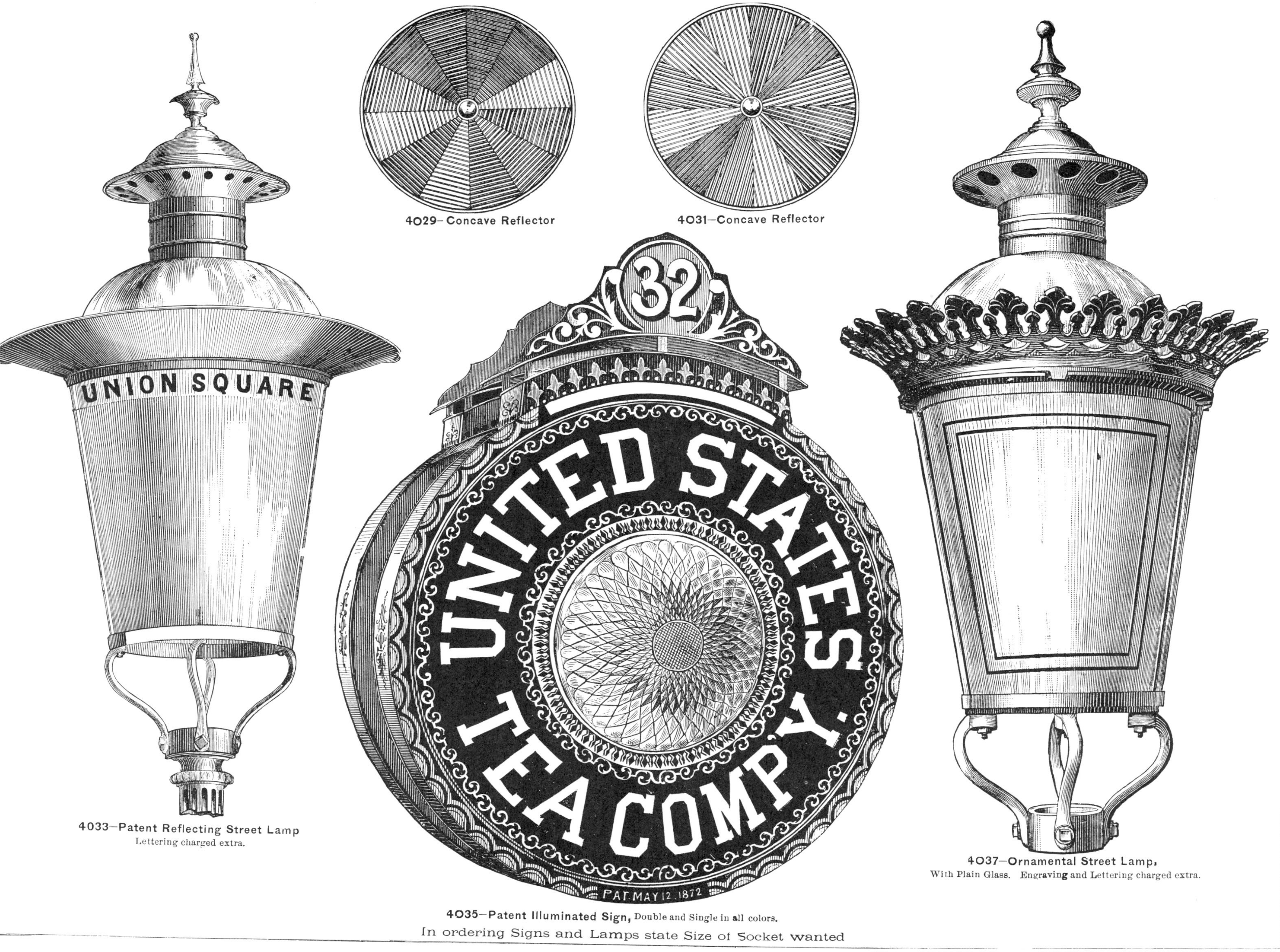

4029–Concave Reflector

4031–Concave Reflector

4033–Patent Reflecting Street Lamp
Lettering charged extra.

4035–Patent Illuminated Sign, Double and Single in all colors.

4037–Ornamental Street Lamp.
With Plain Glass. Engraving and Lettering charged extra.

In ordering Signs and Lamps state Size of Socket wanted